TENSOKU

-Insect preparation-

KEIKI FUKUI | KAZUHO KOBAYASHI

Abbreviation
NHML...The Natural History Museum, London

針で描く芸術

The Art of Drawing with Pins

昆虫標本の、脚や触角、翅などを整えてポージングする仕上げの作業、「展足」。もともとは研究の際、標本を観察しやすくするために行われてきた作業だが、昆虫標本が鑑賞の対象となるにつれ、その仕上げに芸術性が求められるようになってきた。なかでも、日本はこの展足の美しさへのこだわりが特に強い国だ。本書では、そんな世界的に見ても異質な進化を遂げている日本の展足技術を、気鋭の標本作家である福井敬貴氏に紹介してもらった。作業前の状態と、クリーニングや展足を施すことで命が蘇ったかのような姿のコントラストを存分に堪能して欲しい。また、本来であれば人目に触れることのない「針刺し」によってポーズを固定している姿など、その印象的な作業工程にもフォーカスした。無数の針は、実は直接標本に刺さっているのではなく、針を交差させて、その上に脚をのせて持ち上げたり、またはおさえつけたりして固定することで、様々なポーズをデザインしている。まさに針で描く芸術だ。扱っている標本の多くは、大英自然史博物館の歴史的な甲虫コレクションの中から、特に状態の悪いものや造形の面白いものを選んで貸し出してもらったもの。日本の展足技術で、そんな標本たちを蘇らせた。

巻末には、日本では非常に珍しい昆虫標本史の研究者である小林一秀氏に、このような展足文化がどのように生まれ発展してきたのか、その歴史を紹介してもらった。

The meticulous final step in preparing insect specimens, known as spreading or "TENSOKU," involves arranging and posing the legs, antennae, and wings. Originally, this process was carried out to make specimens easier to observe for scientific study. However, as insect specimens gained recognition as objects of admiration, aesthetic considerations became integral to their presentation. Japan, in particular, has developed a unique cultural emphasis on the artistry of tensoku. In this book, Keiki Fukui, a distinguished Japanese specimen artist, introduces the advanced and distinctive Japanese techniques of tensoku, which have evolved in a way unparalleled globally.

The pages invite readers to appreciate the stark contrast between the specimen' s initial state and its transformation following meticulous cleaning and tensoku work, which seems to breathe life back into the insects. Moreover, the book sheds light on the often hidden technique of using pins to set poses—a process that is, in itself, strikingly intricate. Instead of inserting pins directly into the specimens, the artist employs a delicate method of crossing pins beneath the legs and wings, lifting and securing them to shape the ideal pose. It is, indeed, an art drawn with pins.

Many of the specimens featured were selected from the historic beetle specimen collection of the Natural History Museum London, with a focus on those in poorer condition, which were then specially restored using Japanese tensoku techniques.

At the end of the book, Kazuho Kobayashi, a rare specialist in the history of insect specimen, provides insight into the origins and evolution of tensoku culture in Japan, offering readers a deeper understanding of this remarkable craft.

Shogo Kawabata (STRAIGHT)

目次 Contents

Real de
Méx.
H. E. Hinton
Collection.
B.M. 1977-566.

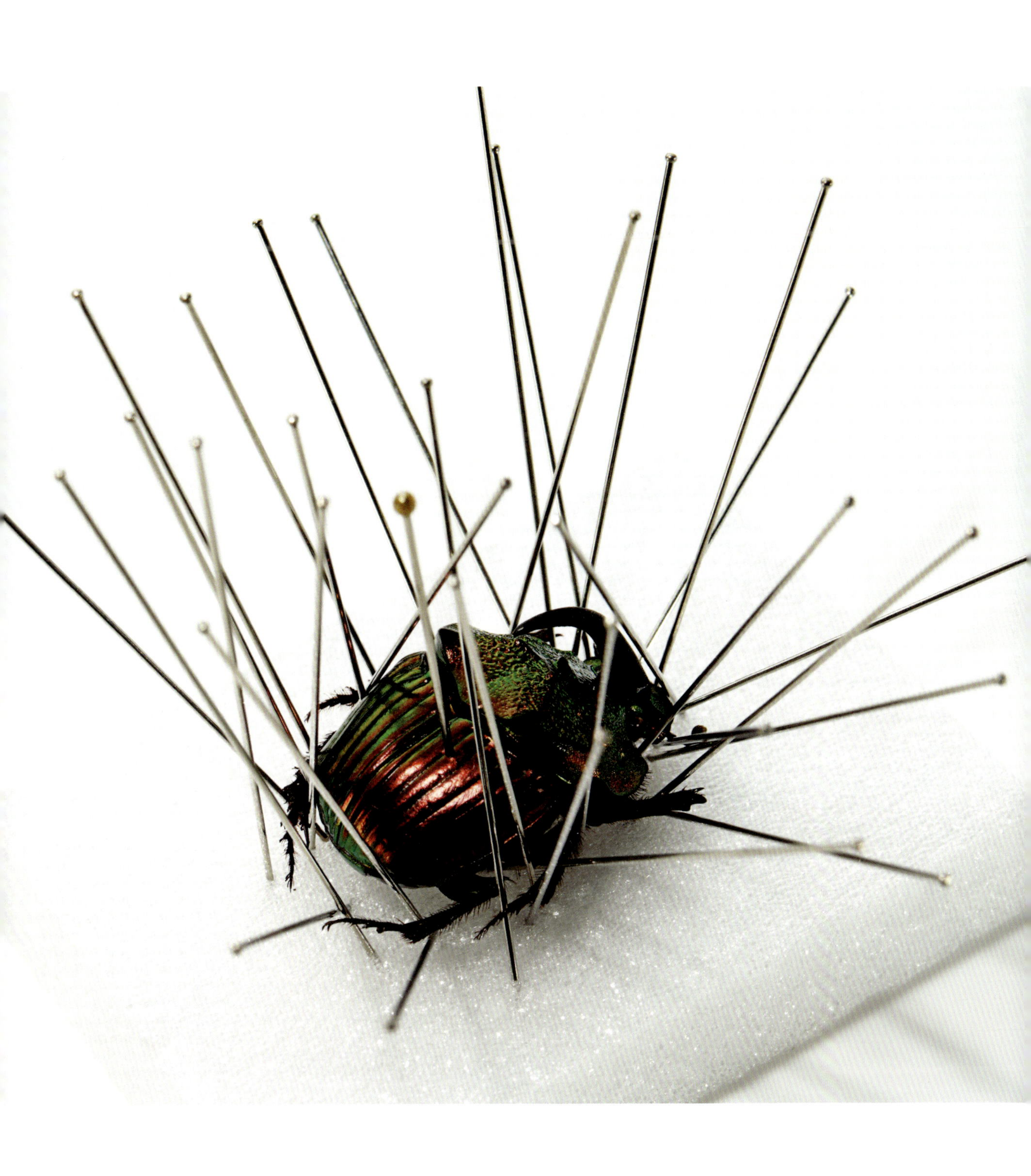

| 展足後 | **AFTER** | TENSOKU | Phanaeus mexicanus (NHML coll.)

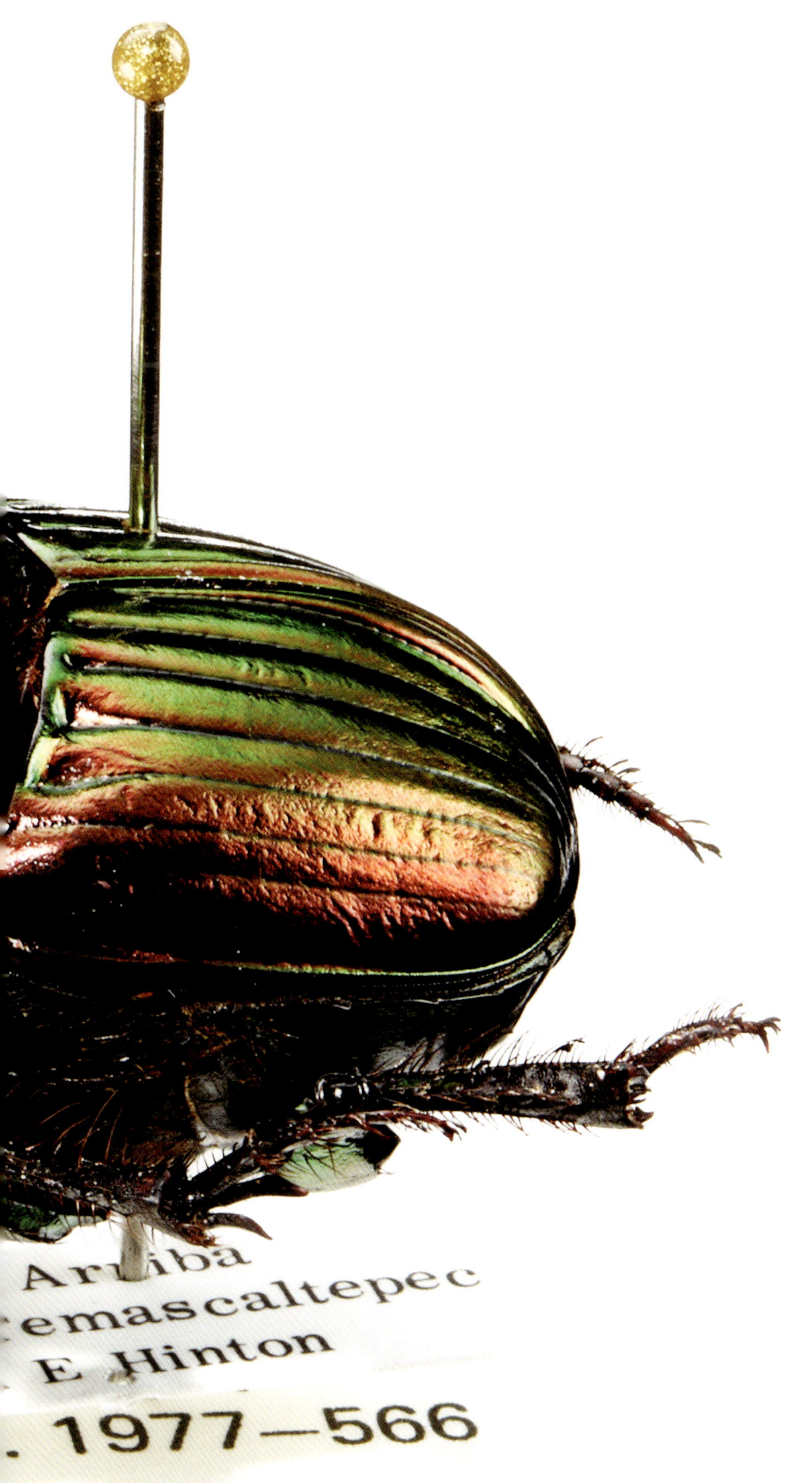
emascaltepec
E Hinton
. 1977—566

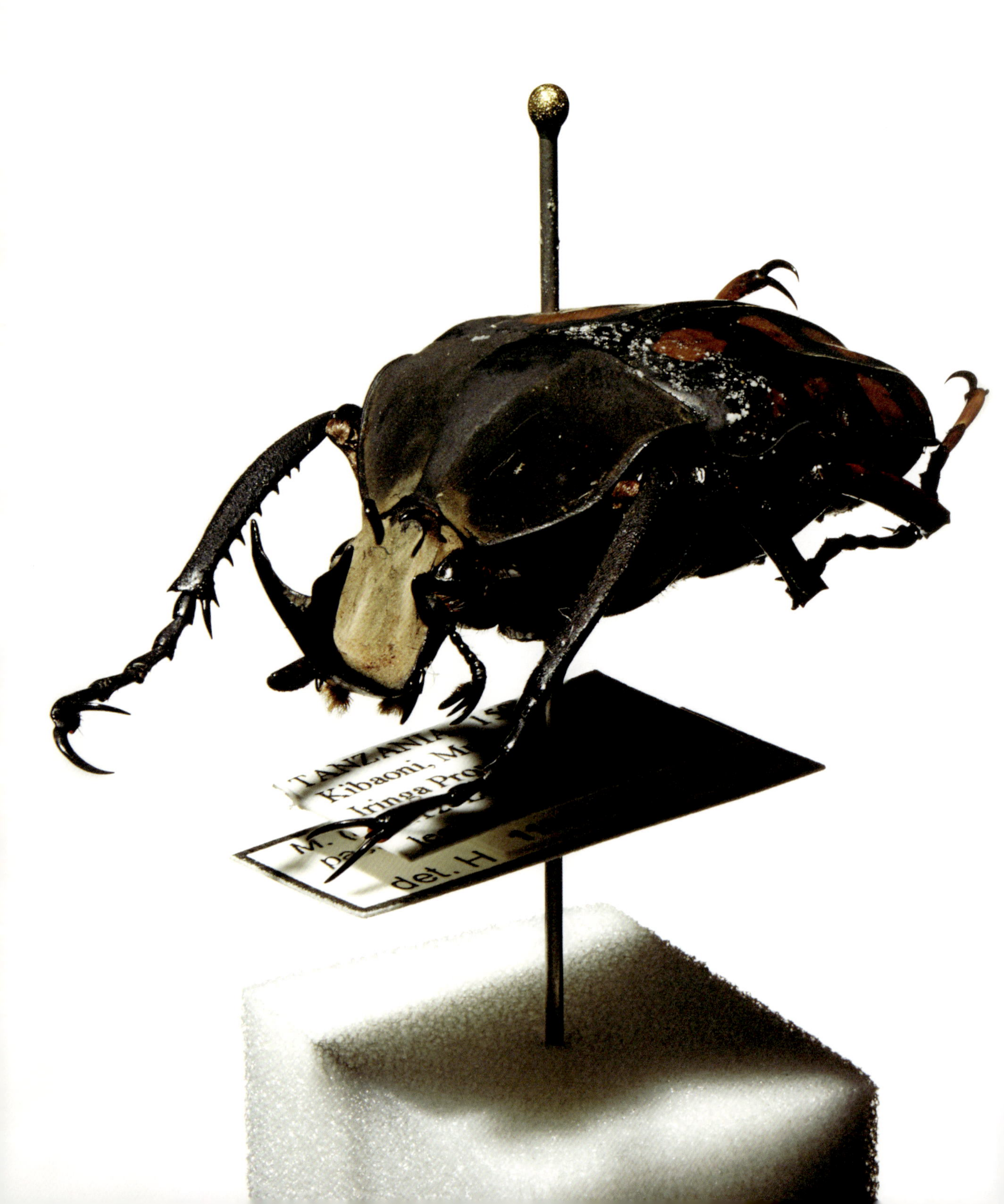
TANZANIA
Kibaoni,
Iringa
det. H

Amaurodes passerinii (NHML coll.) | TENSOKU | **BEFORE** | **展足前** |

| 展足前 | **BEFORE** | TENSOKU | Lucanus cervus (NHML coll.)

| 展 足 後 | AFTER | TENSOKU | Lucanus cervus (NHML coll.)

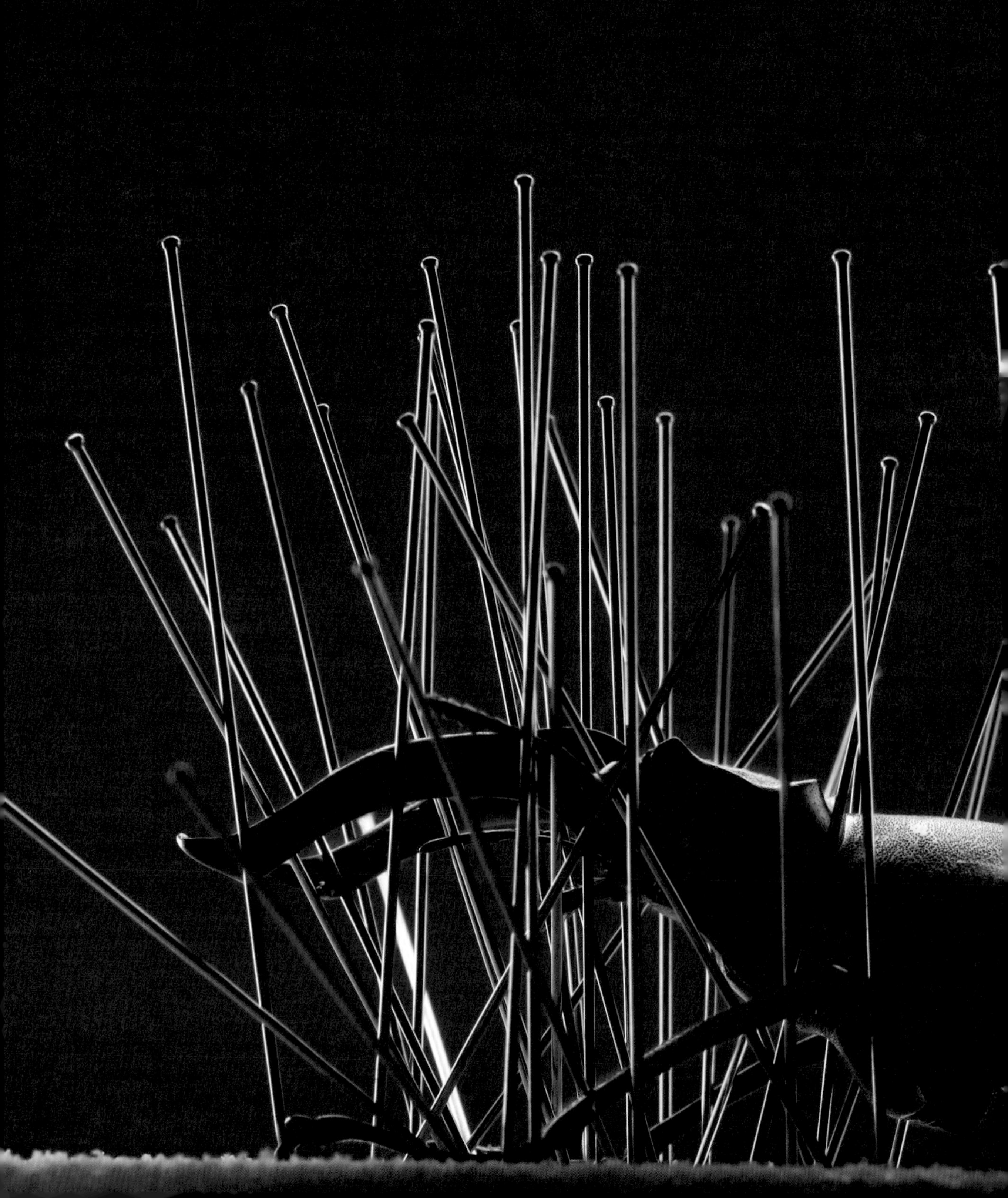

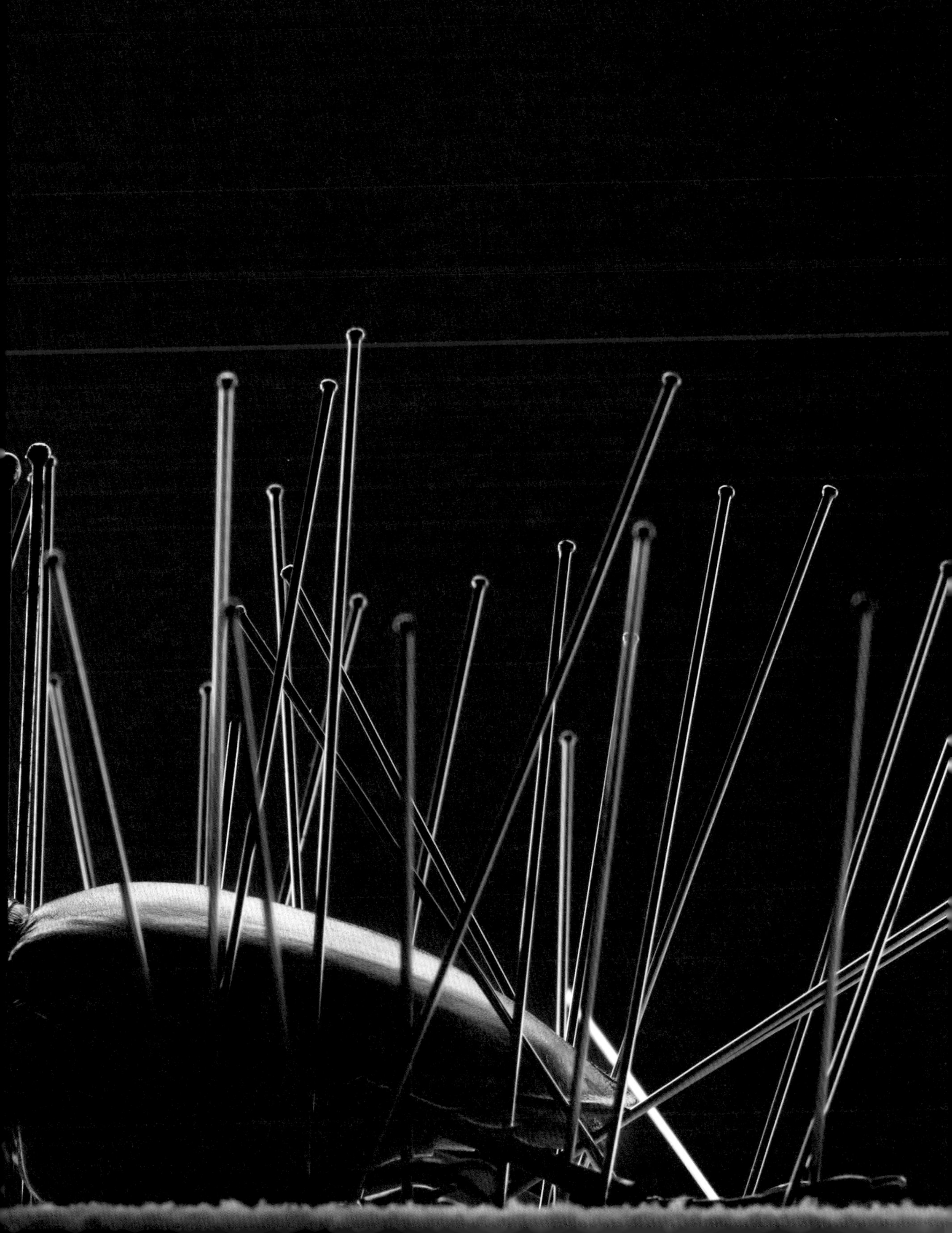

Onymacris bicolor (NHML coll.) | TENSOKU | **BEFORE** | **展足前** |

On sand
dunes
Porto Alexandre
25.ii.1972
B.M.1972-1

Eupholus cinnamomeus (NHML coll.) | TENSOKU | **AFTER** | **展足後** |

| 展足後 | **AFTER** | TENSOKU | Sparrmannia flava (NHML coll.)

| 展足前 | **BEFORE** | TENSOKU | *Rhynchophorus phoenicis* (NHML coll.)

Rhynchophorus phoenicis (NHML coll.) | TENSOKU | **AFTER** | 展足後 |

| 展足中 | **IN PROGRESS** | TENSOKU | Gagatophorus schoenherri (NHML coll.)

| 展足後 | **AFTER** | TENSOKU | Gagatophorus schoenherri (NHML coll.)

N. Luzon

I 展足後 I **AFTER** | TENSOKU | Pachyrhynchus gemmatus (NHML coll.)

Nevinson Coll.
1918—14.

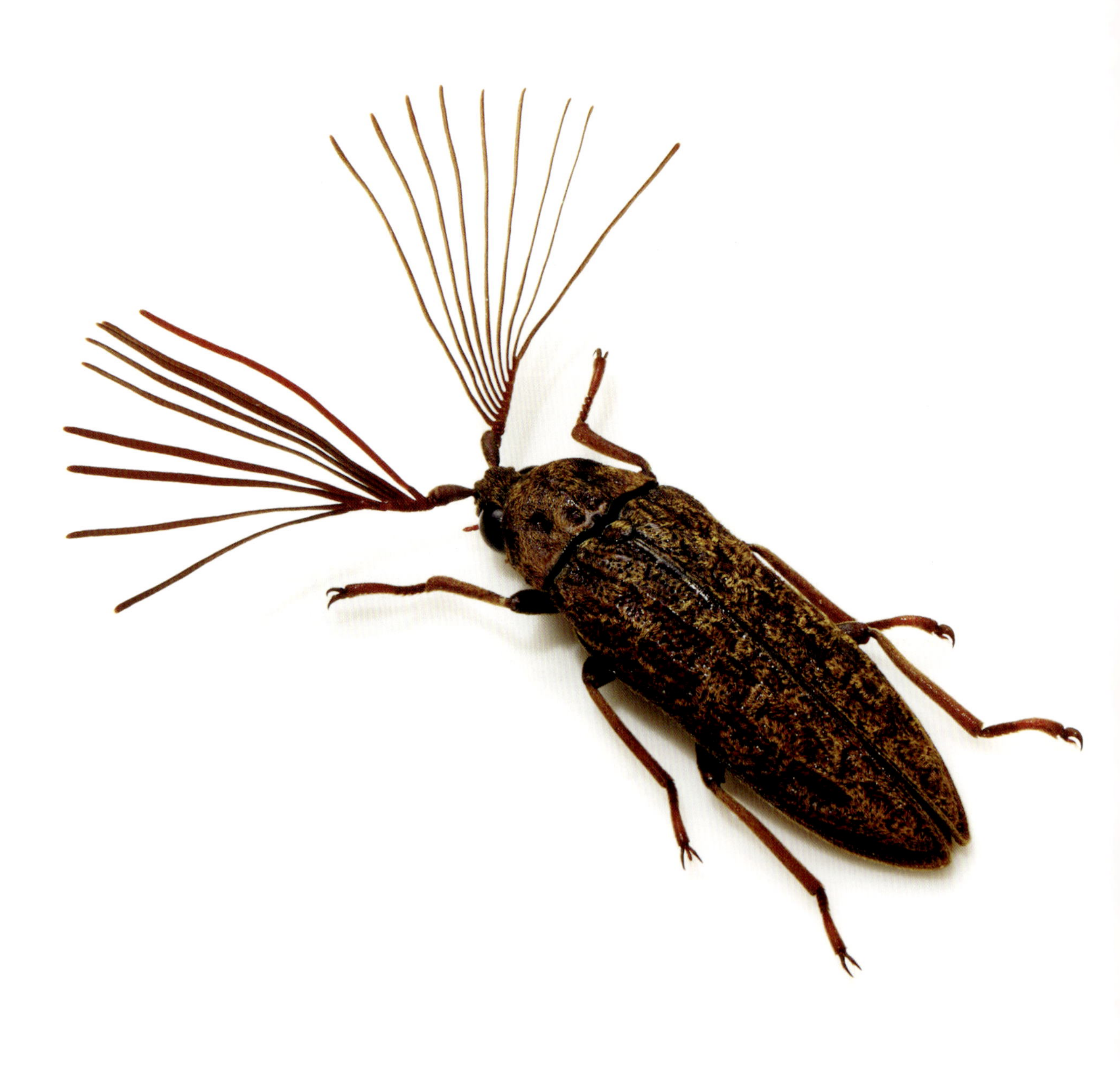

| 展 足 前 | **BEFORE** | TENSOKU | Metallonotus physopterus

| 展足前 | **BEFORE** | TENSOKU | Scarabaeus (pachysoma) bennigseni (NHML coll.)

Scarabaeus (pachysoma) bennigseni (NHML coll.) | TENSOKU | **AFTER** | 展足後 |

PANAMA:_Panama Prov.:
Cerro Jefe
10 MAY 1985

PANAMA: Panama
Cerro Jefe
10 MAY

Cerro

| 展足前 | **BEFORE** | TENSOKU | *Cetoniini* spp.

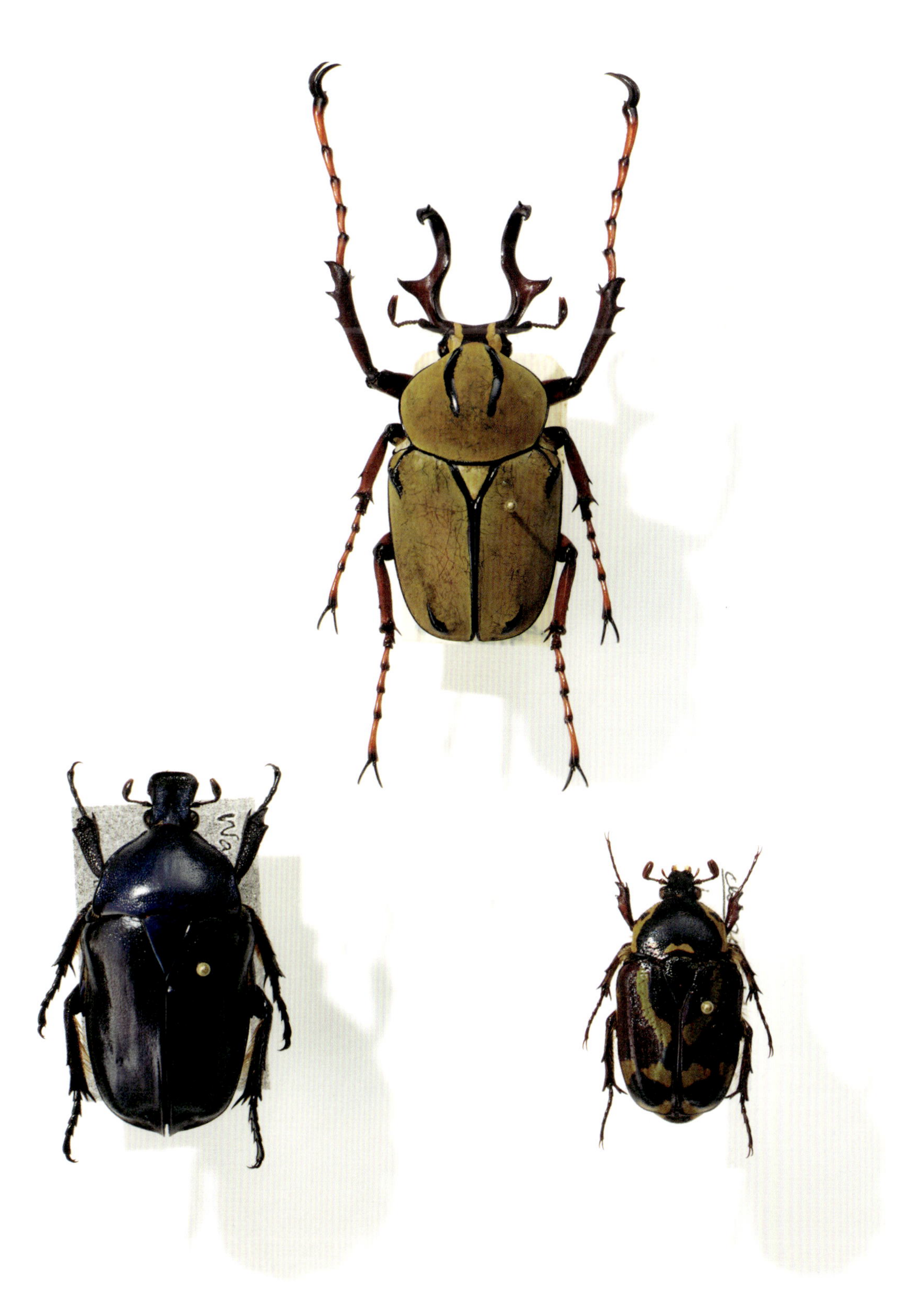

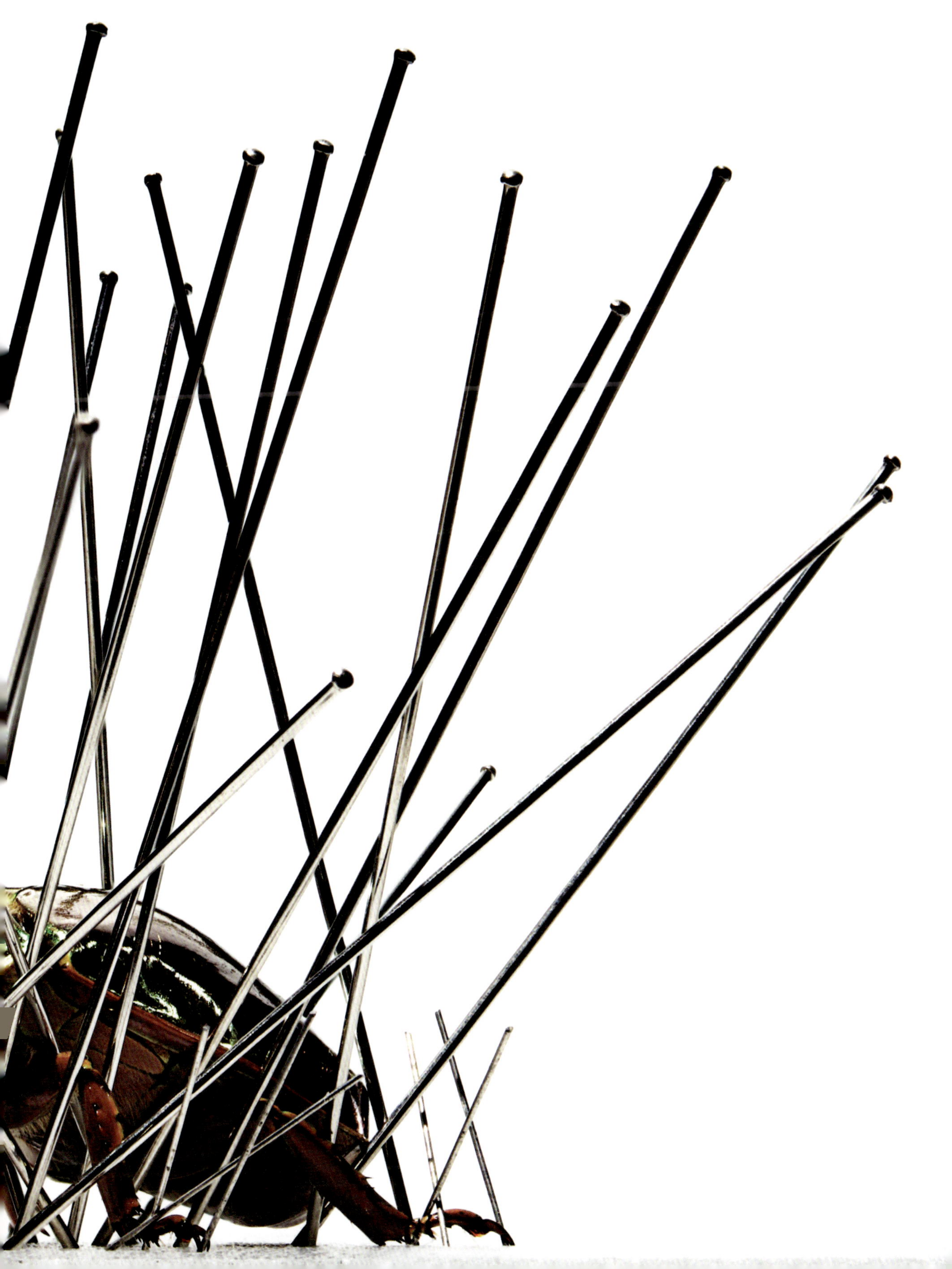

Cyclommatus metallifer

Amaurodes passerinii

Cyclommatus metallifer

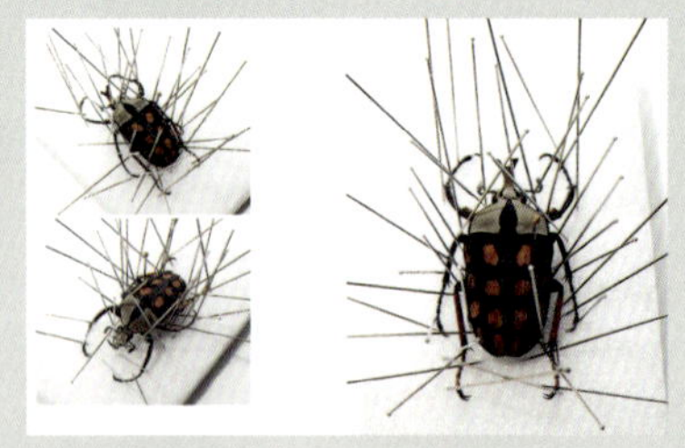

Amaurodes passerinii

Phanaeus mexicanus

Amaurodes passerinii

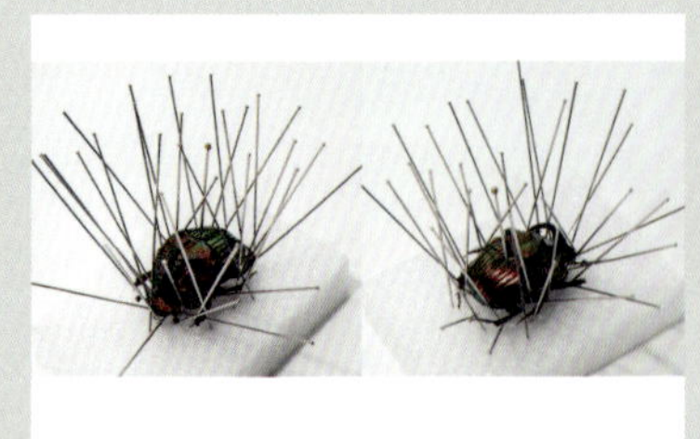

Phanaeus mexicanus

Lucanus cervus

Phanaeus mexicanus

Lucanus cervus

Cyclommatus metallifer

体長の半分が長い大顎で構成されている異形の大型クワガタ。威嚇して大顎を振りかざす姿を立体展足で再現した。

This species is a large Lucanidae characterised by its disproportionately long mandibles, which make up half of its body length. Its aggressive display, with mandibles raised, has been recreated in a three-dimensional tensoku.

Phanaeus mexicanus

草食動物の糞を食べるニジダイコクコガネの仲間。体表に纏わりついた糞を洗い落とすと美しい金属光沢が現れた。(大英自然史博物館所蔵)

This species feeds on herbivore dung. Once the dung that clung to its body was washed away, a beautiful metallic sheen emerged. (NHML coll.)

Amaurodes passerinii

タンザニアの可愛らしい模様をもつツノカナブン。体内の脂分が滲み出てて黒ずんでしまっていたが、脱脂処理により本来の色彩を取り戻した。(大英自然史博物館所蔵)

A charming member of the family Scarabaeidae from Tanzania, this specimen had darkened due to the exudation of internal fats. However, through the process of degreasing, its original colours were restored. (NHML coll.)

Lucanus cervus

イギリス ロンドン産のヨーロッパミヤマクワガタ。この地域の個体群は小型で体高があり、ずんぐりとしている。展足型はヨーロッパスタイルに合わせて製作した。(大英自然史博物館所蔵)

This European stag beetle, sourced from London, UK, exhibits a small and robust phenotype typical of the local population. Its tensoku was crafted in accordance with the European style. (NHML coll.)

Onymacris bicolor

砂漠に生息し、朝霧の中逆立ちして体についた水を飲むキリアツメゴミムシダマシの仲間。脂分で黒ずんでしまった純白の鞘翅を取り戻すクリーニングには苦労させられた。(大英自然史博物館所蔵)

This species inhabits desert environments and is known for drinking water clinging to its body while it is inverted in morning fog. Restoring its pure white elytra through cleaning was quite a challenge. (NHML coll.)

Eupholus cinnamomeus

珍しいブラウンカラーのホウセキゾウムシ。初めてみる種類だったが、大英自然史博物館には大量の本種が収蔵されていた。クリーニングしてみると脚の美しさとコントラストが際立ち、なんておしゃれな虫なのだろうと感銘を受けた。
（大英自然史博物館所蔵）

This rare species of Curculionidae exhibits a striking brown coloration. Although it is extremely rare, the Natural History Museum London houses a large collection of this species. After cleaning, the natural elegance and beauty was highlighted due to the contrasting tones revealed.
(NHML coll.)

Sparrmannia flava

ライオンコガネはナミビアの砂漠に生息するモフモフのコガネムシ。100年以上前の標本なので扱う際は緊張したが、クリーニングしてみると見事な毛並みを取り戻してくれた。
（大英自然史博物館所蔵）

This attractive beetle from the Namibian desert features lion-like tufted hair. Collected over a century ago, handling this specimen was nerve-wracking, but eventually its pristine coat was restored to its original splendor. (NHML coll.)

Rhynchophorus phoenicis

アフリカ産のヤシオオオサゾウムシ。体表に油脂と思われる白い粉が纏わりつき、まるで砂糖菓子のようになってしまっていた。クリーニングしてみると赤色のラインとベルベット調の質感が現れた。（大英自然史博物館所蔵）

This Curculionidae species from Africa had a coating of what appeared to be a white oily powder, giving it the appearance of auger coated sweet. Upon cleaning, red stripes and a velvety texture became apparent. (NHML coll.)

Carabus (Chrysocarabus) spp.

フランスからスペインにかけて生息するコガネオサムシの仲間。まさに歩く宝石だ。オサムシの仲間は脚を縮めた展足がコレクターに人気が高い。

This group of beetle is found from France to Spain and can be described as walking jewels. The Carabidae family is particularly favoured by collectors for their tensoku with retracted legs.

Gagatophorus schoenherri

西オーストラリアに生息するゾウムシ。重厚感のある黒色の体はまるで金属のようだ。しかし、乾燥した標本は非常に軽く重さは1gにも満たない。（大英自然史博物館所蔵）

This Curculionidae species from Western Australia boasts a heavy, metallic black body. However, the desiccated specimen is remarkably lightweight, weighing less than 1 gram.
(NHML coll.)

Lucanus cervus

Rhynchophorus phoenicis

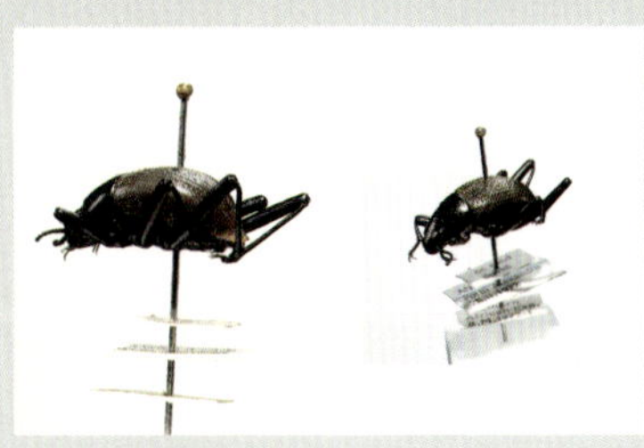

Onymacris bicolor

Carabus (Chrysocarabus) spp.

Onymacris bicolor

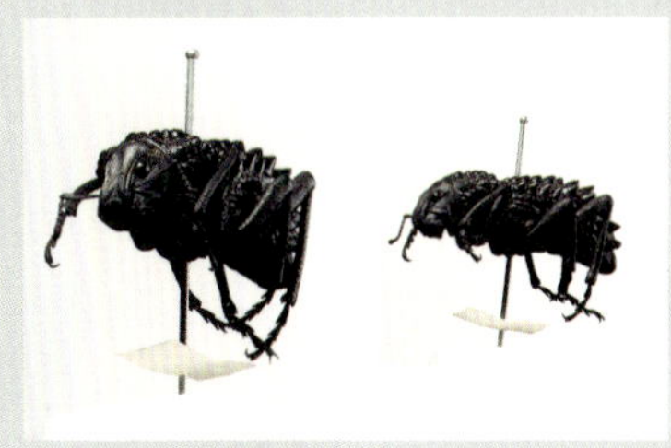

Gagatophorus schoenherri

Eupholus cinnamomeus

Gagatophorus schoenherri

Sparrmannia flava

Gagatophorus schoenherri

Pachyrhynchus gemmatus

Scarabaeus (pachysoma) bennigseni

Pachyrhynchus gemmatus

Scarabaeus (pachysoma) bennigseni

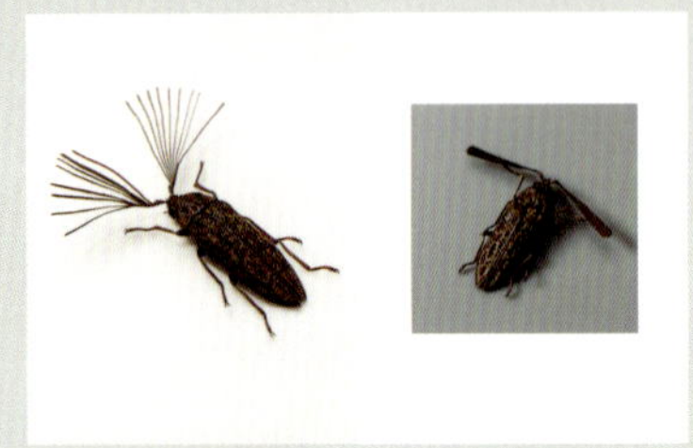

Callirhipis sp.

Cosmisoma ammiralis

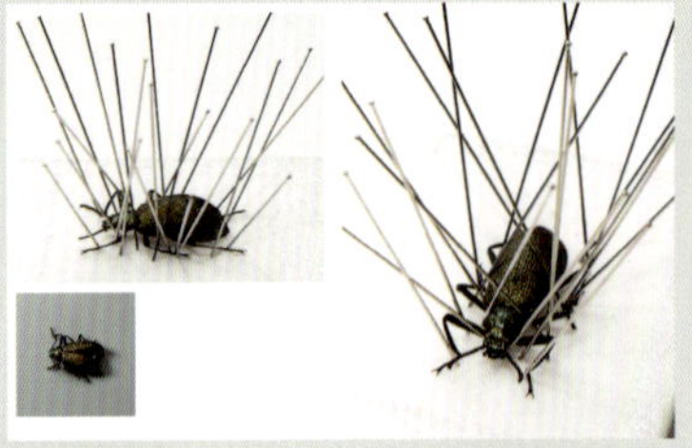

Metallonotus physopterus

Fulcidax bacca

Pachyrhynchus gemmatus

ラメ状の紋を持つカタゾウムシ。欠損が多く、紋のほとんどが視認できなくなっていたが19世紀後半から20世紀前半に活躍した昆虫学者Nevinson Coll. のラベルが付けられた歴史ある個体。100年以上前の標本であるため色が戻るか不安だったが、想像よりもずっと綺麗になってくれた。針が錆びて内部から緑青を発生させていたため、針の刺し替えも行なっている。特定の学術的な研究の際にはあえて針を交換しない場合もある。（大英自然史博物館所蔵）

This Curculionidae is characterised by its glittering, scale-like patterns. Although it had numerous defects that obscured its patterns, this historic specimen bore a label from the Nevinson Collection. Concerns were raised regarding the restoration of its colours due to its age, but it turned out to be much more vibrant than anticipated. The pin had rusted, resulting in the formation of verdigris from within, necessitating their replacement; however, it is not always essential to replace them for academic research purposes.（NHML coll.）

Callirhipis sp.

インドネシア産のナガクシヒゲムシの1種。雄は巨大な触角を持ち、雌を探知する。こうした不思議なパーツを実際に触れて動かし、観察することができるのも展足の醍醐味である。

A species of Callirhipis from Indonesia, where males possess enormous antennae to locate females. The ability to physically touch and manipulate such intriguing features is one of the delights of tensoku.

Metallonotus physopterus

アフリカの乾燥地帯に生息するゴミムシダマシ。鞘翅が開き変形してしまっていたが、適切な軟化と展足を行うことにより修復する事ができた。

A member of the Tenebrionidae from Africa. The elytra had become warped and deformed. However, through appropriate softening and tensoku, it was successfully restored.

Scarabaeus (pachysoma) bennigseni

瓢箪型の飛べないフンコロガシ。糞球を後脚で挟み、前に引きずるように運ぶ珍奇な生態をしている。糞虫の体に付着した糞は標本の情報のひとつとして捉えることも出来るため、落としてしまうか悩みどころ。（大英自然史博物館所蔵）

This flightless dung beetle has a gourd-shaped body and engages in the peculiar behaviour of grasping dung balls with its hind legs and dragging them forward. The dung clinging to the beetle's body can serve as a part of the specimen's information, making the decision of whether to remove it somewhat contentious.（NHML coll.）

Cosmisoma ammiralis

触角にポンポンのような毛束がある南米のカミキリムシ。体色は有毒なベニボタルに擬態していると考えられる。この毛束は何のためにあるのだろうか？不思議な昆虫だ。

This South American longhorn beetle features tufts of hair on its antennae resembling pom-poms. Its body colouration is believed to mimic that of the toxic firefly. The purpose of these tufts remains a mystery.

Fulcidax bacca

ルビーのような輝きを放つ美しいオオコブハムシ。脚や触角は綺麗に畳んで収納できるため、展足前の姿は虫とは思えない。

This beautiful insect exhibits a ruby-like lustre. Its legs and antennae can be neatly folded and stored, making its pre-tensoku appearance rather unrecognisable as an insect.

Dialithus scintillans

パナマの極めて美しい珍品トラハナムグリ。螺鈿細工のような構造色の紋を持ち、角度によってキラキラと光る。左中脚跗節に目立つ修理があったため、再展足の際に修正を行った。

An exceptionally beautiful rarity from Panama, this Cetoniinae species boasts iridescent colouration resembling inlay work, shimmering at different angles. A prominent repair was evident on the left middle leg's tarsus, requiring careful re-adhesion.

Stratioceros princeps

東南アジアに分布する1属1種のカミキリムシ。シックでエレガントな模様をしている。展足前と展足後で随分印象が異なる。

This unique longhorn beetle is found in Southeast Asia and exhibits a chic and elegant pattern. Its appearance changes significantly before and after tensoku.

Cassidini spp.

南米のブローチハムシの仲間。全て同じ産地で見た目も似ているが、なんとこれらは全て別種。属レベルで異なるものが複数含まれている。展足を行うことにより、触角や脚の違いからそれらを見分ける事ができる。

This group of brooch beetles from South America, although similar in appearance and originating from the same locality, comprises entirely different species. Various taxa at the genus level are included. Tensoku enables the differentiation of these taxa based on variations in antennae and legs.

Polybothris sumptuosa

マダガスカルに生息するカワリタマムシ。ダブルピンを用いて飛翔している姿を再現した。普段は鞘翅に隠れている腹部まで美しい。

This species inhabits Madagascar and has been represented in a pose that simulates flight using double pins. Its beautiful abdomen, typically concealed by the elytra, is revealed.

Pseudotorynorrhina japonica (left)
Dicronocephalus wallichii (center)
Chlorobapta frontalis (right)

脂が出て模様が失われたり、汚い状態となってしまっていたハナムグリの仲間たち。脱脂処理と再展足により本来の美しい色味や模様を取り戻した。

This group of Cetoniinae exhibited the loss of pattern due to fat exudation, resulting in an unsightly condition. Through degreasing and re-tensoku, their original, beautiful colours and patterns were regained.

Chrysina optima

オプティマプラチナコガネの立体展足標本。プラチナコガネの仲間は繊細な表面構造を持っており、軟化や乾燥に気を遣う。

Three-Dimensional Tensoku Specimen of Chrysina optima.The members of the genus Chrysina exhibit delicate surface structures, necessitating careful softening and drying processes.

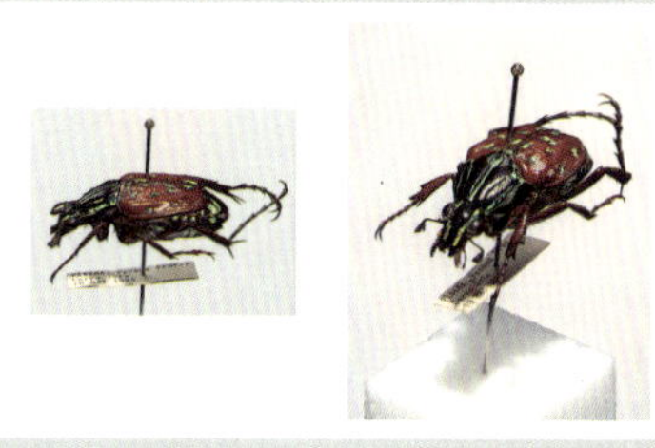

Dialithus scintillans

Cassidini spp.

Dialithus scintillans

Polybothris sumptuosa

Stratioceros princeps

Pseudotorynorrhina japonica (left)
Dicronocephalus wallichii (center)
Chlorobapta frontalis (right)

Stratioceros princeps

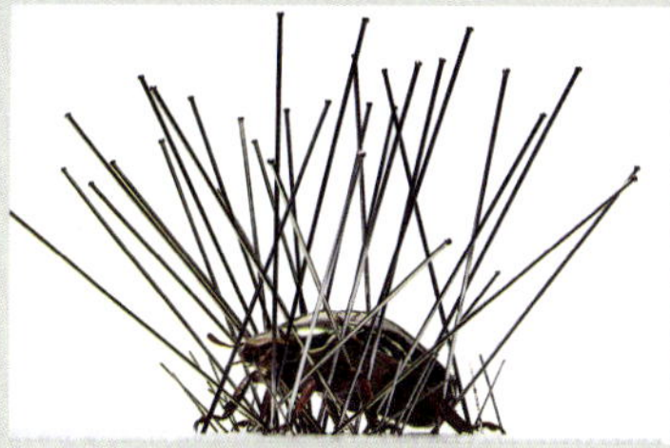

Chrysina optima

Stratioceros princeps

展足技術　福井敬貴

はじめに

展足（てんそく）とは昆虫標本を作成する際、乾燥前に脚や触角を開いて固定して観察しやすくする作業のことです。蝶や蛾など、翅を固定する場合は展翅（てんし）と言います。
日本における昆虫標本の作り方は基本形こそ定義されているものの、その簡易さから、詳しくまとめられた文献は意外と多くありません。特に甲虫類の展足に関しては、作成者個人の技量や美的感覚に委ねられてきました。その結果、日本では展足に「型」や「流派」の様なものが生まれ、独特な文化を形作っています。
本書では主に甲虫類の展足の多様性や文化としての特殊性に着目し、展足という審美的行為がもつ美術・デザイン的な側面を掘り下げ、その技術や制作手法等を紹介します。
私が提示する技法・手法が必ずしも最適であるとは言えませんが、”現代的な展足”として標本製作の際にひとつの参考としていただければ幸いです。

なぜ展足が必要か

実は昆虫標本の作成において、展足は必須な作業ではありません。標本化において最も重要なのはその個体の持つ情報を安全に観察できるように固定し、採集地や年代を記した「データラベル」と共に保存することです。
頑丈な外骨格を持つ昆虫（特に甲虫類）は標本化が容易で、ほとんどの種類は乾燥させるだけでその姿を保存することができます。あとはラベルをつけるだけで立派な標本と呼ぶことができるのです。
では、展足にはどのような意味があるのでしょうか。
優れた展足は標本の観察や比較を容易にし、保存性を高めることが出来ます。展足はデザインと同じで、視覚的な情報量を整え、扱いやすくする技術なのです。また、美しい展足は昆虫の魅力を引き出し、美術品的価値を底上げします。美しい標本は啓蒙普及活動等において昆虫に興味を持ってもらう大きなきっかけにもなり得るでしょう。昆虫は一般的には嫌われものですから、嫌悪感の原因となりうるものを軽減させることは大切なのです。
標本は作成された時点でヒトよりも遥かに長い寿命を持つことになります。綺麗な標本を作ることで、所有者が変わっても大切に保存される標本となるだろうと私は考えています。

型について

展足の基本は“虫の脚・触角を開き、左右対称に整える”というものです。
元々、昆虫標本の蒐集文化は西欧から日本へともたらされました。大英自然史博物館のヒストリカル・コレクションを見てみると、名称はなくとも、標本を整えるという行為は博物学黎明期から存在していたことが見受けられます。
しかし現代において、ヨーロッパと日本の展足はそれぞれ独自の発展を遂げています。
ヨーロッパ式は保存性を保ちながら標本を多視点から観察するものとして捉えており、やや立体的な展足型となっている傾向があります。それに対し、日本式は1枚絵で特徴が全て捉えられるような真俯瞰構図からの写像性が強く意識されていて、脚を持ち上げた平面的な展足が好まれます。
日本は昆虫図鑑の発行が盛んなため、おそらく図鑑のプレート化などの撮影の際に、より説明的な画像が求められていった結果ではないかと私は考察しています。また、ヨーロッパと比べ湿度が高くカビや虫害が起きやすい日本では、標本箱の開閉にリスクが伴います。なるべく箱に収めた状態のまま観察を行う工夫であると捉えらることもできるでしょう。

Introduction

Tensoku is a Japanese term, both noun and verb, referring to the process of spreading and fixing the legs and antennae of an insect specimen before drying, making it easier to observe.

Although the basic technique for preparing insect specimens in Japan is widely recognised, detailed literature on the subject is surprisingly scarce due to its simplicity. Particularly in the case of Coleoptera, the process has largely been left to the individual skills and aesthetic judgment of the preparator. This has led to the development of distinct “styles” or “schools” of tensoku, which now form part of Japan's unique cultural tradition. In this book, we focus mainly on the diversity of tensoku in Coleoptera and its special cultural significance.
I delve into the artistic and design aspects of this aesthetic practice, introducing its techniques and methods. While we cannot claim that the techniques we present are universally optimal, we hope they can serve as a reference for what could be considered “modern tensoku” in specimen preparation.

The Necessity of Tensoku

Strictly speaking, tensoku is not an essen-

TENSOKU techniques

Keiki Fukui

ヨーロッパ式

European Style

跗節を横に寝かせて底面まで下げ、脚は内向きで全体的に丸みを帯びた立体的な形に整えられる。 大型甲虫の場合、迫力を出すために虫体の下に板を挟んで展足し体高を高くする場合もあるようだ。 脚が比較的コンパクトにまとめられているので移動の際に破損しにくく、標本箱に並べた際のデッドスペースが生まれにくい。脚に角度がついているため、画像における脚の長さなどの比較検討にはあまり適さない。

In this style, the tarsi are laid horizontally on the pinning board, with the legs directed inward, creating a rounded, three-dimensional shape. For larger Coleoptera, an additional board may be placed underneath the body to increase its height for a more imposing appearance. Since the legs are arranged compactly, this style minimises damage during transport and reduces dead space in the specimen box. However, as the leg tips are angled downward, this style is less suitable for comparative studies of leg length when photographing the specimen from directly above.

European Style

Japanese Style

日本式

Japanese Style

跗節を立て、脚は外側に引っ張ったように直線的で、横から見ると平行になるように上に浮かせて固定される。画像における比較検討に長けるが、ツメなどが引っかかりやすく移動時の破損率がやや高い面がある。また、脚を伸ばすため標本箱に並べた際にスペースを大きく使ってしまう欠点がある。平面的で、写像性を意識した展足といえる。

In this style, the claws are positioned vertically relative to the pinning board, with the legs splayed outward. When viewed from the side, the legs align parallel to the elytra's lateral surfaces, and the femora are lifted to give the specimen a flatter appearance. This configuration is well-suited for comparative imaging, but the claws tend to catch on the pinning board when removing the specimen, increasing the risk of damage. Additionally, the extended legs take up more space in the specimen box, limiting the number of specimens that can be stored in one box. The Japanese style is thus more planar and focused on imaging accuracy compared to the European approach.

tial step in preparing insect specimens. The primary goal when producing an entomological specimen is to ensure that the information intrinsic to the individual can be safely observed and preserved along with its "label data" detailing location and date of collection. Insects with sturdy exoskeletons, especially Coleoptera, are relatively easy to prepare, as most species can simply be dried to preserve their form. A specimen labelled as such can then be considered complete.

So, what is the purpose of tensoku? A well-executed tensoku facilitates observation and comparison of specimens while also enhancing preservation. Similar to design, tensoku is a technique that organises visual information to make the specimen easier to handle. Additionally, beautiful tensoku accentuates the appeal of insects, adding to their aesthetic value as works of art. A finely prepared specimen can inspire interest in insects for public outreach and education, serving as an entry point for many. Since insects are often disliked, it is important to mitigate elements that might provoke aversion.

Once a specimen is produced, it is likely to outlive its producer by a significant margin. We believe that crafting a specimen with care will ensure that it is treasured and preserved, even as it changes hands over time.

On Styles

The basic principle of tensoku is to "spread the legs and antennae symmetrically". The culture of collecting insect specimens was originally introduced to Japan from the West. Historical collections at the Natural History Museum, London indicate that, while it might not have been explicitly named, the practice of arranging specimens existed before the dawn of natural history.

In contemporary times, tensoku in Europe and Japan has evolved independently. The European-style tensoku tends to focus on durability and multi-angle observation, resulting in a more three-dimensional presentation. In contrast, Japanese-style tensoku emphasises capturing all key characteristics from a single top-down perspective, creating a flattened effect where the legs appear raised when viewed from the side. It is thought that this distinctive Japanese approach developed due to the prevalence of insect field guides with photographs, which required more explanatory images for publication. Furthermore, Japan's relatively humid climate, which promotes mould and pest activity, may have encouraged observation of specimens without removing them from the specimen box.

クワガタムシの展足型

Tensoku Styles in Lucanidae

クワガタムシは日本の蒐集家に特に人気があり、コレクターによって名称が付けられている場合もあります。ここでは例として、私が影響を受けた5つの展足型を主観込みで再現しました。

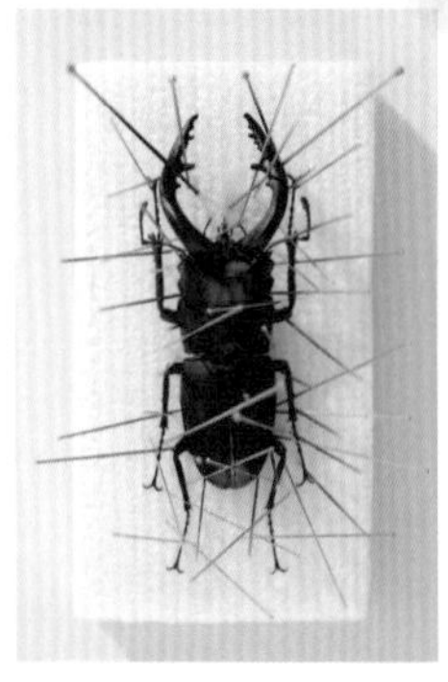

鈴村型
Suzumura style

関西の有名コレクター鈴村克彦氏の型。前脚は上方に、後脚は下方に大きく引き伸ばしたような型で、ツメを大きく開き、大顎はやや閉じ気味。

This style, from renowned Kansai collector Katsuhiko Suzumura, features forward-extending forelegs and rearward-extending hind legs. The claws are spread wide, and the mandibles are slightly closed.

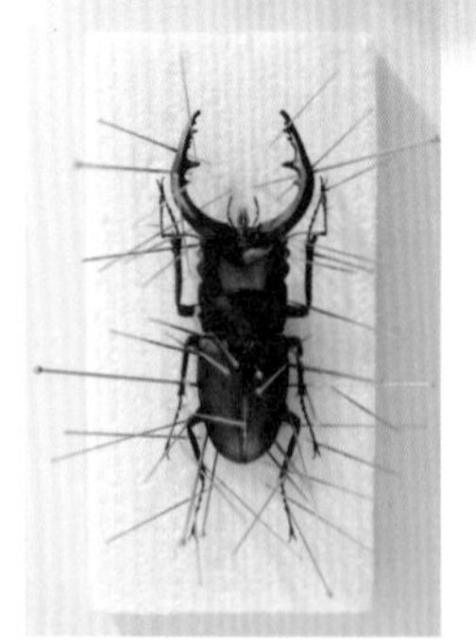

藤田型
Fujita style

有限会社むし社会長・藤田宏氏の型。跗節はリラックスした角度で固定され、ツメは30～40度ほど開き、大顎を大きく開く。

This style, by Hiroshi Fujita, chairman of Mushi-sha Co., features tarsi at relaxed angles, claws opened about 30–40 degrees, and wide-open mandibles.

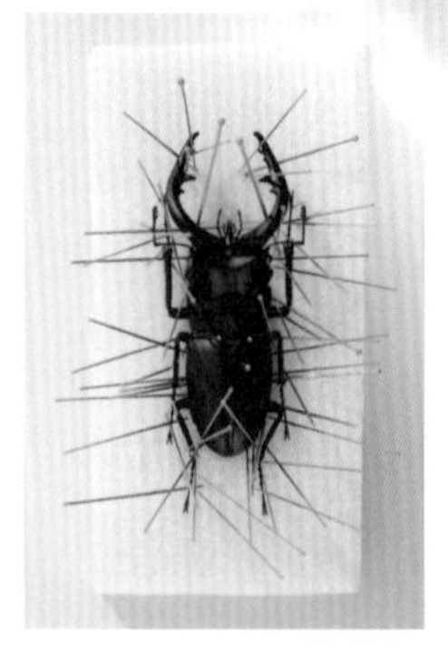

前田型
Maeda style

株式会社インセクエスト代表・前田健氏の型。頭部・前脚がやや上向きで、前脚・中脚跗節が脛節から直線的に固定される。大顎はやや開き、ツメは閉じ気味。

Created by Ken Maeda, president of Insequest Co., this style has the head and forelegs angled slightly forward. The fore- and mid-leg tarsi are set in a straight line from the tibiae, with slightly open mandibles and nearly closed claws.

Lucanidae is especially popular among Japanese collectors, with some tensoku styles named after the collectors themselves. Here, I have recreated five tensoku styles that have influenced me.

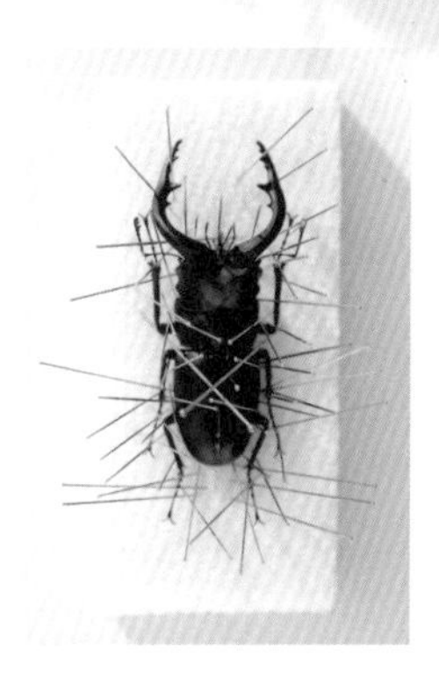

Jinlabo 型

Jinlabo style

ミヤマクワガタ類の研究家・佐藤仁氏の型。後脚がやや上方に持ちあげられ、跗節は全体的にやや外側を向く。ツメは大きく開く。

Developed by Lucanidae researcher Jin Sato, this style has hind legs slightly raised forward and tarsi opened outward. The claws are spread wide.

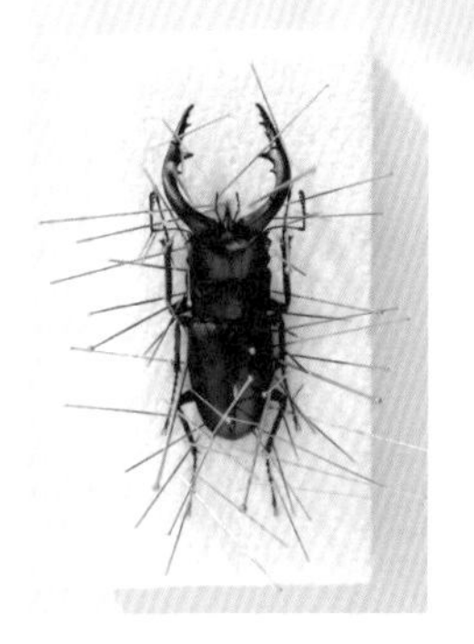

八戸型

Hachinohe style

八戸インセクト研究会の市川裕二氏、吉川克哉氏らの型。前脚は下方に引き下げられ、前脚・中脚跗節は脛節から外向きに伸ばされている。ツメは閉じ気味。

Created by Yuji Ichikawa and Katsuya Yoshikawa of the Hachinohe Insect Research Group, this style features rearward-drawn forelegs and outwardly spread fore- and mid-leg tarsi from the tibiae. Claws remain mostly closed.

オサムシの展足型

Tensoku Styles in Carabidae

オサムシの仲間は脚や触角をコンパクトに畳んだ展足が主流となっています。ヨーロッパ由来の型であると言われています。この型はオサムシの魅力である宝石のような身体を強調し、繊細な触角や跗節の破損を防ぐことができます。しかし、近年は脚を畳んでいることに違和感を覚え、脚を開いた型に展足する人も増えている様です。

In the Carabidae family, it is common to fold the legs and antennae compactly. This style, thought to have originated in Europe, highlights the jewel-like bodies of these beetles while preventing damage to their delicate antennae and tarsi. However, there has recently been a trend toward spreading the legs in tensoku due to aesthetic preferences.

優れた展足型とは？

Ideal Tensoku Styles

展足型には個人の美意識や考え方が反映されています。そこにケチをつけたりするのはあまりにも野暮です。しかし、優れた標本とはどの様なものなのか?ということは標本を作ったことがある人なら誰しもが1度は考えることでしょう。
そこで今回は、私個人の展足に対する考え方を紹介させていただきます。
展足とは情報量の整理、記号化であると私は考えています。膨大な情報量をもつ標本を視覚的に記号化することで、標本が持つ特徴をより捉えやすくするというものです。また、保存や管理のし易さも重要です。写像性を求めた説明的な展足は管理を行う標本箱内に無駄なスペースを生み出しやすいことが問題となってきます。
そこで私は、『展足を最小限の長方形構図に収める』ことによりその解決を図っています。
標本箱の規格サイズは基本的に長方形です。その中に標本を収めるのなら、長方形に虫を展足することで無駄なスペースを作らずに済みます。また、構図の範囲内ならば脚を大きめに広げても管理に支障をきたすこと少なく、台紙への収まりも良いです。

Tensoku styles reflect personal aesthetic sensibilities and philosophies, and it would be unrefined to criticise them. However, anyone who has prepared specimens will likely ponder what constitutes an excellent specimen. Here, I share my perspective on tensoku. Tensoku organises and symbolises information. By visualising the vast information within a specimen, its defining characteristics become more accessible. Ease of storage and management is also essential. When prioritising explanatory tensoku styles, it can be challenging to avoid unused space within a specimen box. To address this, we aim to arrange specimens within a minimal rectangular composition, fitting them compactly within the standard rectangular box dimensions. Even with legs spread, this configuration facilitates management without hindrance and ensures neat placement on the mounting board.

立体展足

Three-Dimensional Tensoku

生体時の姿を再現した、剥製的な側面が強い展足です。ライブ標本や動態標本とも呼ばれますが、私は平面的な展足と対照的であることから親しみを込めて立体展足と呼んでいます。
基本的に通常の標本箱には収めることが難しいため、作製時は専用の容器も作ることが大切です。

This tensoku style emulates a taxidermic approach, reproducing the insect's natural posture. As such specimens are challenging to fit within regular specimen boxes, specialised containers are recommended.

標本作成技法

道具類　Tools

ピンセット　Tweezers

最も重要な道具。数種類を用意し、用途や状況によって使い分ける。先端は砥石を使って研ぎ、バネは弱くなるように調節する。先端に溝があるものも用意しておくと大型種の修理などに便利。

The most essential tool; it is helpful to have several types to use according to needs. Sharpen the tips on a whetstone and adjust the spring tension. Tweezers with grooved tips are useful for repairing larger specimens.

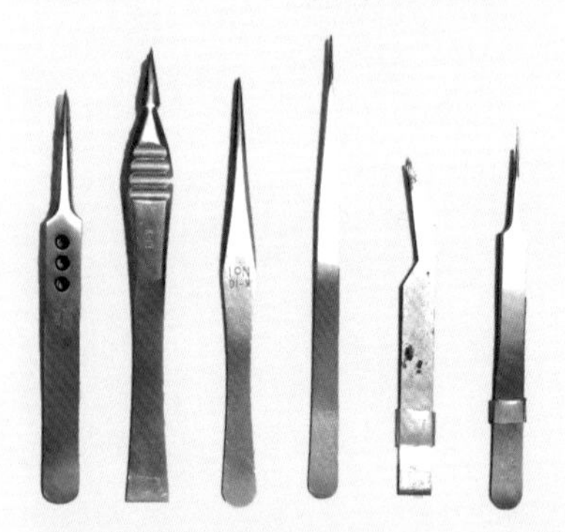

実体双眼顕微鏡／拡大鏡

Stereo Microscope and Magnifying Glass

小型種の展足や修理に使用。倍率は4倍～20倍程度。私は極小のターンテーブルをセットで使用することが多い。

Used for tensoku or repairing smaller specimens, with a magnification of 4x to 20x. A small turntable can also be helpful.

平均台　Pinning Block

標本の高さを整える際に使用する。製品によって若干規格が異なるため注意。

Used to adjust the height of specimens. Note that standards may vary slightly between products.

展足板 Pinning Boards

適度な硬さの発泡スチロール板を使用する。断熱材のスタイロフォームやカネライトフォームは入手が容易で安価なため重宝する。

Polystyrene boards of appropriate hardness are used. Insulation materials like Styrofoam and Kanenite foam are readily available, inexpensive, and thus highly convenient for this purpose.

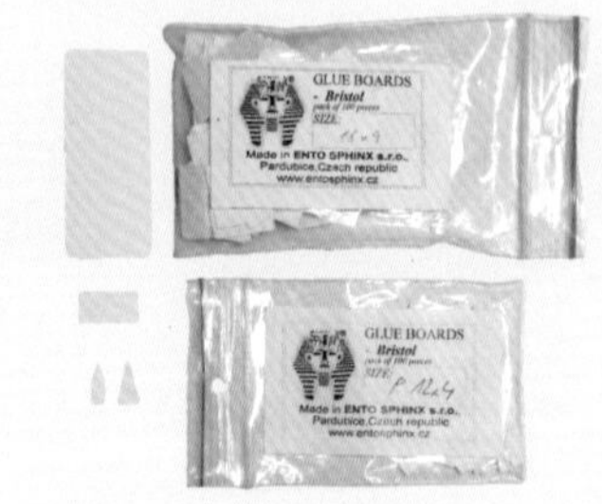

台紙　Mounting Boards

大小様々なサイズがあり、チェコ製やドイツ製のものが専門店で販売されている。虫の大きさに合わせて適したサイズを使用する。

Available in various sizes, with Czech- and German-made options available from specialist suppliers. Choose an appropriate size based on the insect's dimensions.

標本針　Pins

必ずステンレス製のものを使用する。私は展足には志賀昆虫針1～3号、虫や台紙に刺す針はチェコ製のナイロンヘッド昆虫針を使用している。ナイロンヘッド昆虫針は針の表面がわずかにザラついており、標本が回転しにくいためである。

ヨーロッパでよく使用されるブラックエナメル針は湿度の高い日本では錆が発生しやすいため不向き。だが、針が硬いため展足には重宝する。7号針や微針も用意しておくと良い。

Use only stainless steel pins. For tensoku, Shiga Entomology Supply Co.'s pins No. 1–3 are recommended, with European nylon-head pins for piercing specimens to the elytra or mounting board. European black enamel pins are prone to rust in Japan's humid climate, making them unsuitable for piercing specimens but helpful for tensoku due to their sturdiness. It's also advisable to have No. 7 pins and minuten pins available.

接着剤　Adhesives

木工用ボンド、膠を状況によって使い分ける。ボンドは適度に水で薄めたものを用意しておくと便利。修理には低粘度の瞬間接着剤などを用いる場合もある。必ず水や溶剤で剥離可能なものを使用する。

Use wood glue or animal glue as required. Diluted wood glue, is convenient to have prepared in advance, thinned to the consistency required with water.
Low-viscosity cyanoacrylate glue may also be used for repairs. Ensure that the adhesives used can be removed with water or solvent.

標本の入手とその後の処理 Obtaining and Processing Specimens

標本は自身で生息地に赴いて採集する他、専門店や展示即売会で購入することも出来ます。最初の処理はその後の標本作製に大きく影響するので、自己採集の場合は丁寧な仕事を心がけたいところです。

Specimens can be obtained either by collecting them directly from their natural habitats or by purchasing them from specialist shops or insect fairs.

採集の場合

採集した昆虫は酢酸エチルで殺虫・防腐処理を行うのが一般的。自然死だと既に腐敗が進んでいたり、脚に癖がついてしまって製作が困難となる場合がある。乾燥後に色が残りにくい一部の昆虫はこの際に亜硫酸ガスに適度な時間あてることで色彩の保存が容易になる。

〆た直後の昆虫は死後硬直を起こしているため、脚や大顎を無理に動かすと破損に繋がる。チャック付きポリ袋等にいれて乾燥を防ぎつつ硬直が解けるのを待ち、正しく関節が動くか確認してからタトウ紙に並べて乾燥させる。

Collected Specimens

For collected insects, it is customary to use ethyl acetate for insecticide and preservation. Specimens that have died naturally may already show signs of decay, or their muscles may be dried and hardened, making preparation challenging. For certain insects prone to discolouration upon drying, exposing them to sulphurous acid gas for a controlled period can help prevent this. Freshly euthanised insects undergo rigor mortis; forcing movement of the legs or mandibles can cause damage. Place them in a sealed plastic bag to prevent drying and wait until rigor has dissipated, confirming that the joints move freely.

購入の場合

販売されている標本は大抵の場合パッキング処理され乾燥している。標本の内部や脱脂綿のなかにカツオブシムシやシバンムシ等の標本を食べる虫が潜んでいることがあるため、購入後は一度冷凍して殺虫処理を行うことをおすすめする。標本はタッパーなどの密閉容器で防虫剤と共に保管をする。

Purchased Dry Specimens

Purchased specimens are usually packed and dried. As they may harbour specimen-eating insects such as Dermestidae or Ptinidae inside their bodies or cotton wool, it is recommended to freeze the specimens for a few days post-purchase for pest control. Store specimens in airtight containers, such as tupperware, with insect repellent for long-term preservation.

標本作製手順 Specimen Preparation Procedure

①脱脂

Degreasing

甲虫標本の管理において最も問題になるものの一つが、虫の体に含まれる油脂成分が体表面へと染み出してきてしまう事象です。これは越冬を行う寿命の長い種類や、肉食性が強い種類などでよく見られる特徴ですが、ほぼ全ての甲虫はいずれ脂が出るといっても過言ではありません。
脂が出た虫は体全体が油分で汚れ、水分量で体表の色が変化する種は黒ずみ、構造色で発色している種は本来の発色を失ってしまうなど、個体独自の視覚的情報を大きく損なってしまいます。
また、油分には軟化作用があり展足が崩れてしまう他、台紙から剥落し標本が破損する大きな原因にもなります。これらを予防するため、脂が出ているものは展足前に必ず脱脂処理を行い、そうでないものも念のため処理を行っておくことが好ましいです。
また、カビが発生してしまった標本は内部がカビの菌糸に犯されているため、脱脂と同様の薬品処理を行って殺菌することをおすすめします。

One of the main challenges in maintaining Coleoptera specimens is the leaching of body oils onto the surface over time. This phenomenon is especially prevalent in species with long lifespans or carnivorous habits, though nearly all Coleoptera may eventually secrete oils. When this occurs, the specimen's surface becomes greasy, and for species whose colouration depends on moisture, the surface may darken or lose its original structural colours, significantly diminishing its visual qualities. Additionally, the softening effect of the oils can cause tensoku to collapse, and the specimen may detach from the mounting board, leading to damage. To prevent these issues, it is essential to degrease specimens showing oil secretion before tensoku, and it is advisable to treat others as a precaution. For mould-affected specimens, it is recommended to use chemical treatments similar to those for degreasing to disinfect the internal hyphae.

アセトン、ベンジンA、キシレン等の揮発性有機溶剤を用いた方法

タッパー等の溶剤に耐性のある密閉容器に虫が浸る程度の量を注ぎ、漬け込むことによって脱脂を行う。虫の大きさに合わせた小型容器を用意し、虫の体高程度の量を注いで複数回溶剤の交換を繰り返して脱脂すると比較的短時間で綺麗に脱脂を行うことが出来る。漬け込む期間は虫の状態や大きさに左右され、ひどく脂が出ているものほど時間がかかる。また、気温が低いと時間がかかるため18度以上の環境にて管理することが望ましい。大抵の場合は小型のもので6時間～12時間、中型のもので12～24時間、大型のもので24～60時間程度。巨大種や状態によっては1週間以上かかる場合もある。しかし基本的には数をこなし、状態を見極める慣れと勘を身につける必要がある。

脱脂が進むと液体の色が透明から黄色へと変わり、レモンティー程度の色になったら溶剤の交換を行う。液体の色に大きな変化が見られなくなったら溶剤より引き上げ、溶剤を完全に揮発させてから軟化・クリーニング等を行う。脱脂後の標本は限りなく脱水されており極めて破損しやすいため、扱いに注意する。

また、これらは引火性が高く健康への有害性を持つ有機溶剤であるため、使用後の溶剤は必ず火元のない場所で厳重に管理し、地域の処分方法に従って廃棄を行う必要がある。

Method Using Acetone, Hexane, Xylene, or Other Volatile Organic Solvents

Fill a solvent-resistant, airtight container with enough solvent to submerge the insect, then soak it to achieve degreasing. A small container matched to the insect's size allows efficient solvent use and facilitates multiple solvent changes for faster degreasing. The required soaking time varies depending on the specimen's size and condition; heavily oiled specimens take longer, and low temperatures will extend the process, so maintaining a temperature above 18 ° C is preferable. Generally, small specimens require 6–12 hours, medium-sized specimens 12–24 hours, and larger ones 24–60 hours. Some large species or heavily oiled specimens may take over a week. Judging the necessary time requires experience in assessing the condition of each specimen. As degreasing progresses, the solvent changes from clear to yellow; when it reaches a tea-like colour, replace the solvent. Once the liquid no longer changes colour, remove the specimen, allowing the solvent to fully evaporate before softening and cleaning. Specimens degreased in this way are extremely dehydrated and fragile, requiring careful handling. As these solvents are highly flammable and hazardous to health, solvent must be stored away from open flames and disposed of according to local regulations.

酸素系漂白剤を用いる方法

適度に濃度を調節した酸素系漂白剤に標本を漬け込み、脱脂する手法。標本への影響が大きいため小型種や繊細な種への使用はおすすめしない。脱色作用があるため、翅の染み汚れなどの変色にも効果がある。但し、長期間漬け込みすぎると脱色が進み触角や跗節の末端から赤くなってしまうため注意が必要。また、タンパク質の保存状態が良い標本はカタラーゼ反応により腹部が膨張して膨らむ場合がある。これを利用して、乾燥すると縮んでしまうような柔らかな昆虫を膨らませて保存する技法も存在する。標本が極めて脆くなってしまうため、間違っても塩素系漂白剤を使用してはならない。

Method Using Oxygen-based Bleach

Soak the specimen in a suitably diluted oxygen-based bleach solution for degreasing. This method is not recommended for small or delicate specimens due to its strong effects. The decolourising properties can also treat discoloured areas, such as wing of bleach. However, prolonged exposure may cause fading, turning the antennae and tarsi red from the tips, so caution is required. Specimens with well-preserved proteins may experience abdominal swelling due to catalase reactions. This effect can be used as a technique to restore the natural shape of soft insects that would otherwise shrink upon drying. Chlorine-based bleach should never be used, as it makes specimens extremely brittle.

② 軟化

Softening

乾燥した標本は非常に脆く繊細ですが、水分を含ませて軟化を行うことで脚や触角を安全に動かすことができる様になります。軟化した標本は全ての関節がしっかりと動くかストレッチを行い、確認してからクリーニング・展足作業へと移ることが重要です。

Dried specimens are extremely brittle and delicate, but by softening them with moisture, the legs and antennae can be safely adjusted. After softening, it is important to stretch all the joints to ensure they move smoothly before proceeding to cleaning and tensoku.

軟化に注意が必要な昆虫

プラチナコガネは急激な湿度変化を与えるとヒビが入ってしまう場合がある。フトタマムシの仲間やゴミムシダマシの一部は、水溶性の粉状物質を纏っている場合があるので、極力ぬらさず、熱を与えないなど工夫が必要。

Insects Requiring Special Care During Softening

For Chrysina (Platinum Beetle), sudden changes in humidity may cause cracks on its pronotum and elytra. Certain Sternocera (Buprestidae, Jewel Beetle) and Tenebrionidae (Darkling Beetles) species may be covered in water-soluble powdery substances, so special care is needed to keep them dry and avoid heat.

湿潤軟化

すのこ付きタッパーなどの容器に水を浸した脱脂綿やティッシュペーパーなどを敷いた湿潤器を作り、そこに標本を並べ空気中の湿度から自然に水分が吸収されるのを待つ軟化方法。
フトタマムシ類など濡らすと体表の分泌物が流れ落ちたり染みができたりする恐れがあるもの、発色が色素由来の昆虫や特殊な構造色を持つプラチナコガネなど、繊細な昆虫にはこの方法が適している。
大型甲虫になるほど軟化に長い時間を要するため、巨大な甲虫には不向き。また、体表の塵等はそのまま残ってしまう欠点がある。

Moist Softening

This method involves creating a humidifier in a container, such as plastic box, with dampened cotton wool or tissue paper, and placing the specimens inside to allow them to gradually absorb moisture from the air. This technique is suitable for delicate insects like Sternocera, which may develop smears or discolouration if the body surface is exposed to water, as well as for Chrysina, whose colours arise from structural pigments. Large Coleoptera require longer softening times, so this method may not be suitable for particularly large beetles. Dust and debris on the body surface remain an issue with this technique.

浸漬軟化

水を容器に標本が浸かる程度注ぎ、標本を浸して軟化する手法。複数の個体を同時に軟化する場合、容器の底に脱脂綿等を敷いておくと接触による破損を防ぐことが出来る。2%程度に希釈したオスバン液を使用するとクリーニング効果が得られ軟化も速い。繊細な小型の甲虫に適している。

Immersion Softening

This technique involves immersing the specimen in water within a container until it softens. If softening multiple specimens simultaneously, placing cotton wool at the bottom of the container can prevent damage from contact. Using a solution diluted to around 2% with benzalkonium chloride (osvan solution) enhances cleaning effects and speeds up softening. This method is suitable for delicate small Coleoptera.

湯浸軟化

耐熱性のある容器に湯を注ぎ、標本を浸して軟化する手法。大型の甲虫によく用いられる。体表の油分、塵や汚れを浮き上がらせるため高いクリーニング効果が期待できる。オスバン液や中性洗剤を添加したものを用いると尚良い。短時間で軟化させることが可能なため、クワガタムシ、大型のカミキリムシなど筋肉が強く頑丈な甲虫に適している。水分を多く虫体に含ませることから、展足後の乾燥には長い期間が必要。

Warm Water Immersion Softening

In this method, the specimen is immersed in warm water in a heat-resistant container. Often used for larger Coleoptera, it provides effective cleaning by lifting oils, dust, and debris from the body surface. Using a solution with added benzalkonium chloride solution or a neutral detergent improves results. This method allows rapid softening, making it ideal for strong, muscular Coleoptera such as Lucanidae and large Cerambycidae. Since insects absorb significant moisture, they require a longer drying period post-tensoku.

浸漬軟化／湯浸軟化
Immersion Softening / Warm Water Immersion Softening

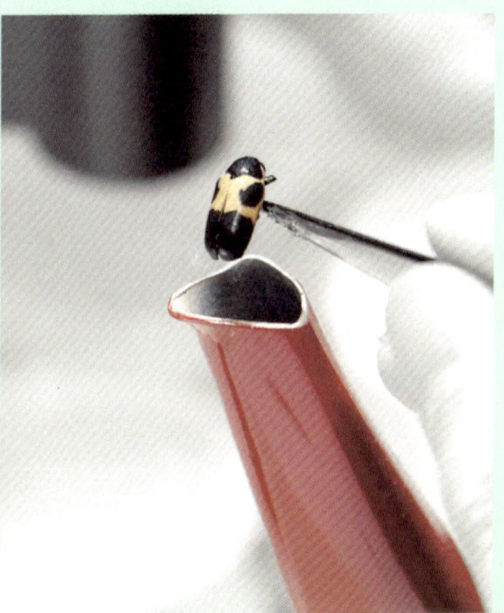

蒸気軟化　Steam Softening

蒸気軟化

電気ポット等で沸騰させた湯の水蒸気に標本を当てるようにして軟化する手法。
虫のサイズによるが、小型の甲虫であれば大抵数十秒程度で軟化できる。給湯器にザルを組み合わせた蒸気軟化装置も考案されており、これを用いれば大型の甲虫も短時間で余分な水分を含ませずに軟化を行うことが出来る。
非常に素早く軟化を行うことが可能であり、標本に含まれる水分量が少ないため乾燥が早い。最もスピーディーに標本制作を行うことができる軟化方法と言える。
欠点としては、熱による変質があるものには使用できないこと、クリーニング効果は期待できない他、乾燥の早さから素早く展足を行う必要性があり、虫体構造のしっかりとした理解が必要となる。そのため、やや玄人向きの手法であると言える。蒸気軟化は部分的な軟化を可能とするため、乾燥中の標本の跗節や触角の修正などにも用いることが出来る。

Steam Softening

This technique involves softening the specimen with steam from boiling water, such as from an electric kettle. For small Coleoptera, softening is usually achievable within a few seconds. A specialised steam softening device, which combines a hot water dispenser and a sieve, allows efficient softening of large Coleoptera without excessive moisture absorption. Steam softening is remarkably fast and allows for quick drying due to minimal moisture retention, making it an efficient softening method for specimen preparation.

However, it is unsuitable for heat-sensitive specimens and provides little cleaning effect. Due to the rapid drying, tensoku must be performed promptly, which requires a thorough understanding of insect structure. Thus, steam softening is considered a more advanced technique. This method also enables selective softening, which is useful for adjusting tarsi or antennae during the drying process.

その他の軟化技法
筋肉破壊 / 軟化剤

死後硬直後そのまま乾燥してしまったもの、不適切な薬品で処理されたものなどは前脚基部、大顎などの筋肉が引きつったまま固まってしまい軟化を行っても動かせないほど硬い場合や、癖がついていて展足が難しい場合がある。こうした標本は“ゴム化”と呼ばれており、無理に動かすことは破損の可能性が非常に高く厄介。
安全に動かすためには関節膜をピンセット等で貫き、内部の筋肉を適度に破壊する必要がある。筋肉破壊を行う際は、しっかりとした軟化を事前に行うことが重要。また、タンパク質分解酵素等の軟化剤を必要な箇所付近の関節膜から注射器を用いて注入することによって、ゴム化を解消出来る場合もある。大型甲虫は軟化剤の効き目が薄いことも多いため、筋肉破壊と併用して行うと良い。

Other Softening Techniques
Muscle Destruction / Softening Agents

Specimens that dried without proper preservation, or that were treated with unsuitable chemicals, may exhibit a rigid stiffness, particularly in the foreleg base or mandibles, due to contracted muscles. Known as “rubberisation,” these specimens are extremely difficult to move without risking damage. To safely adjust them, the joint membrane should be pierced with tweezers to partially destroy the internal muscles. Adequate softening beforehand is essential for muscle destruction. Injecting a softening agent, such as a protease, through the joint membrane can also sometimes remedy rubberisation. For large Coleoptera, a combination of muscle destruction and softening agents is often more effective.

③クリーニング
Cleaning

軟化を行い柔軟性・強度がある状態のうちにクリーニングを行います。クリーニングを怠ると再度軟化から標本を作り直す必要性が生じる場合も。虫の体表には必ずと言って良いほど塵や汚れが纏わり付いています。汚れは標本の持つ情報のひとつであるとも言えますが、観察や撮影の際に大きな障害となる他、カビなどが発生する原因となる場合もあります。数種の筆や柔らかい歯ブラシなどを用意しておくと便利です。

Cleaning should be carried out while the specimen is still softened and flexible. Neglecting this step may require restarting the preparation process with another round of softening. Dust and dirt are almost always present on the insect's body surface. Although these contaminants can be considered part of the specimen's information, they can obstruct observation and photography and may promote mould growth. Having various brushes and a soft toothbrush is convenient for cleaning.

アルカリ電解水、希釈したオスバン液・中性洗剤等

霧吹きを使って標本に吹き付け、表面の汚れを溶かし落とす手法。オサムシや糞虫などの油分やタンパク質系の汚れに効果的。体表に残留する場合があるためクリーニング後は真水や精製水で洗浄を行う。

*アルカリ電解水はプラチナコガネなど、一部の昆虫の発色に影響を及ぼす場合があるため注意。

Alkaline Electrolysed Water, Diluted benzalkonium chloride Solution, Neutral Detergent, etc.

This method involves spraying the specimen to dissolve and remove surface contaminants. It is effective for removing oil and protein-based dirt, especially from Carabidae and Scarabaeinae (dung beetles). Since residues may remain on the surface, specimens should be rinsed with pure or distilled water after cleaning. Note: Alkaline electrolysed water may affect the pigmentation of certain species, such as Chrysina.

ボンドパック／水のりパック

接着剤を体表に薄く塗り、乾燥後剥がすことによってまとめて塵を取り去る手法。体表に深い溝があり、しつこい泥汚れが多い種などに非常に有効。但し、体表に繊細な毛を備えるもの、体表が鱗毛や分泌物で覆われているものには使用してはならない。

Wood glue Pack / Water Glue Pack

This technique involves applying a thin layer of adhesive to the body surface and peeling it off after drying to remove dust in bulk. It is highly effective for species with deep grooves or heavy mud deposits. However, it should not be used on specimens with fine body hairs, scales, or secretion-covered surfaces.

エアダスター

風圧で汚れを吹き飛ばす手法。私は塗装用ハンドピースをコンプレッサーの圧力を調節しながら使用している。寝てしまった毛を立てたり、余分な水分を取り除くのにも極めて有効。

Air Duster

This technique uses compressed air to blow off dirt and dust. It is especially useful for lifting flattened hairs and removing excess moisture. An airbrush for painting can also be used with controlled air pressure for precision.

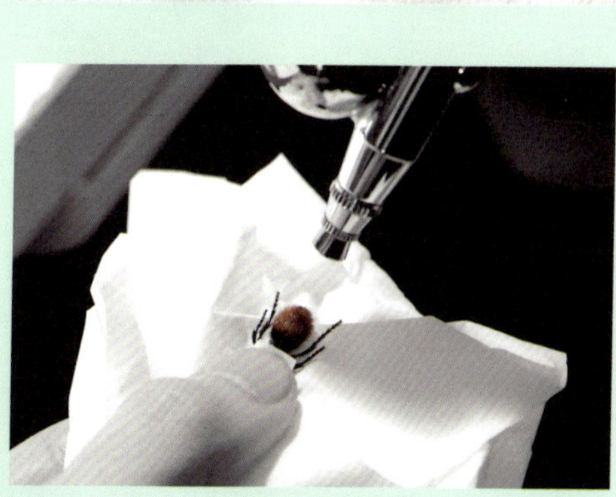

④展足［針展足］

Tensoku in large-sized species

展足の固定方法は人それぞれのやり方があり、型や標本の状態によっても異なるものですが、ここでは私の日本式の展足方法を1例として紹介します。

The method of tensoku varies depending on individual preferences, tensoku style, and the condition of the specimen. Here, I will introduce one example of my Japanese-style tensoku method.

軟化・クリーニング後は全ての関節が問題なく動くかストレッチを行う。この際、ある程度脚や触角を開いておくと展足作業に入りやすい。

After softening and cleaning, check the mobility of all segments by stretching them. At this stage, slightly extending the legs and antennae will facilitate the tensoku process.

虫を固定しやすくするため、展足板を虫の体の大きさに合わせて溝を彫る。掘を作らず、針のみで体を固定する場合もある。展足板に線を引きガイドラインを作るのも効果的である。

To stabilise the insect, carve a groove on the pinning board to match the size of its body. In some cases, the body can be secured using only pins without carving a groove. Adding gridlines to the pinning board is also very useful as a guide.

右鞘翅から後胸腹板へ針を垂直に刺す。一気に貫通させず、途中で位置を確認しながら刺すと良い。中脚基節を貫かないよう注意。

Insert the pin vertically through the right elytron and into the metaster-num. Avoid piercing through in one go; instead, check the position midway through. Be careful not to pierce the base of the middle leg.

体を針で押さえつけるように固定し、脚をピンセットで動かして腿節→脛節→跗節の順に針で固定していく。針はX状に交差させ、展足板に押さえつけるようにすると固定しやすい。

Secure the body by pressing it down with the pin, then use tweezers to adjust the legs, fixing each segment—starting from the femur, then the tibia, and finally the tarsus—using pins. Cross the pins in an "X" shape to press the legs against the pinning board, making them easier to stabilise.

触角の固定は極めて重要で、分類学的に重要なだけでなく視覚的な印象にも大きく関わる。固定は脚より後でも先でも問題ないが、乾燥が進むと破損しやすいため注意。無理に固定せず、後に部分軟化を行い修正するのも良い。

It is crucial to spread the antennae, as they are important not only for taxonomic observation but also for aesthetic purposes. Antennae can be pinned before or after the legs, but care should be taken as they become fragile as drying progresses. If they do not spread easily, it is possible to partially soften them later using steam and adjust them.

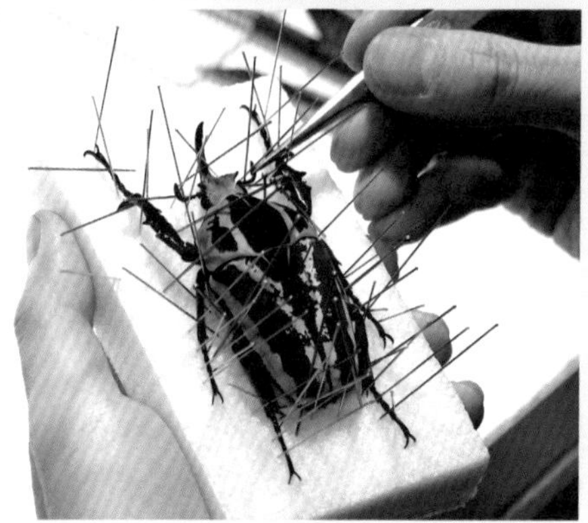

展足後は湿度の低い場所でしっかりと乾燥させる。軟化具合にもよるが中型種で数日、大型種で数週間程度。衣装ケース等に乾燥剤を敷いた乾燥BOXを作るのも良い。ただし、急激な乾燥は鞘翅の割れや開きなどを招く場合があるため注意が必要。

After tensoku, thoroughly dry the specimen in a low-humidity environment. The drying time varies depending on the degree of softening—several days for medium-sized species and up to several weeks for larger species. Using a drying box, such as a storage container lined with desiccants, is recommended. However, be cautious, as rapid drying may cause cracks or separation in the elytral suture.

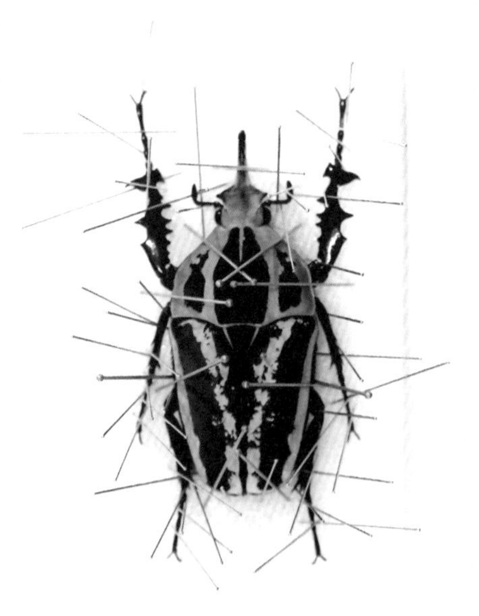

⑤展足［テープ展足］
Tensoku in small-sized species

小型種は両面テープ等を用いることで針をあまり使わずに展足することができます。タマムシやハムシ、ゾウムシなど、跗節が幅広い虫に特に有効です。小型ハチ類などの展翅展足にも応用可能。絆創膏や付箋の糊面などを使う方法も考案されています。

For small species, tensoku can be done with minimal use of pins by employing double-sided tape. This method is particularly effective for species with broad tarsi, such as those in the Buprestidae, Chrysomelidae, and Curculionidae families. It can also be applied to small Hymenoptera species. Methods using adhesive surfaces from plasters or sticky notes have also been devised.

展足板はスチロール板を虫の大きさに合わせてカットすると扱い易い。両面テープも虫の大きさや強度に応じて粘度の違うものを使用する。

Cutting a Styrofoam board to match the insect's size makes it easier to handle as a pinning board. Choose double-sided tape with an adhesive strength suited to the size and strength of the insect.

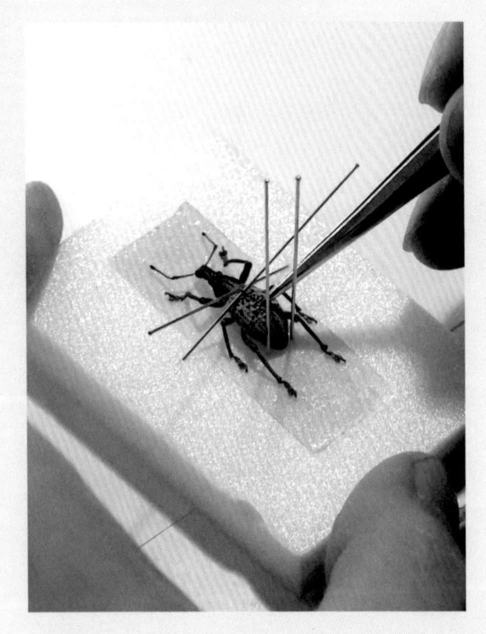

体を針で固定し、脚・触角をピンセットでテープに貼り付けて固定していく。ピンセットでの調整がメインとなるので、自身の手に合うピンセットを調節することが重要。

Secure the body with a pin, then use tweezers to attach the legs and antennae to the tape. Adjustments are mainly done with tweezers, so it is essential to set up a comfortable workspace for ease of handling.

固定に微針等を使用する場合もあるが、状態の良い標本であれば軟化具合をコントロールすることでほとんど針を使わずに展足できる。
この個体は体を固定する針は途中で外してしまった。

In some cases, minuten pins may be used for stabilisation. However, if the specimen is in good condition, tensoku can be achieved with minimal use of pins by carefully controlling the degree of softening. For this specimen, I removed the pins that initially secured the body partway through the process.

10mm 以下の小型種は顕微鏡や拡大鏡の下で展足を行う。
私は小型ターンテーブルにマスキングテープの粘着面が上になるように貼り付けて使用しており、ほとんどの場合針は使わずピンセットのみで展足している。

For small specimens under 10 mm, tensoku is performed under a microscope or magnifying glass. I use a small turntable with masking tape (adhesive side up), and in most cases, I perform tensoku using only tweezers without pins.

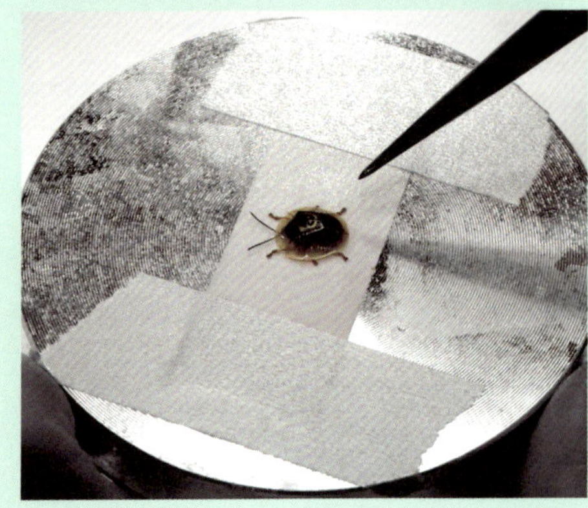

小型種は乾燥が早いため数時間～数日程度の乾燥でマウント可能となる場合が多い。両面テープから剥がしにくい場合は無水エタノールを面相筆で流し込むと容易に剥がすことができる。
また、糊が付いてしまった針はアセトンに漬け込むことで綺麗になる。

Since small species dry quickly, they can often be mounted after drying for only a few hours to a few days. If the specimen is difficult to remove from the double-sided tape, applying anhydrous ethanol with a fine brush will facilitate removal. If adhesive remains on the pins, soaking them in acetone will clean them.

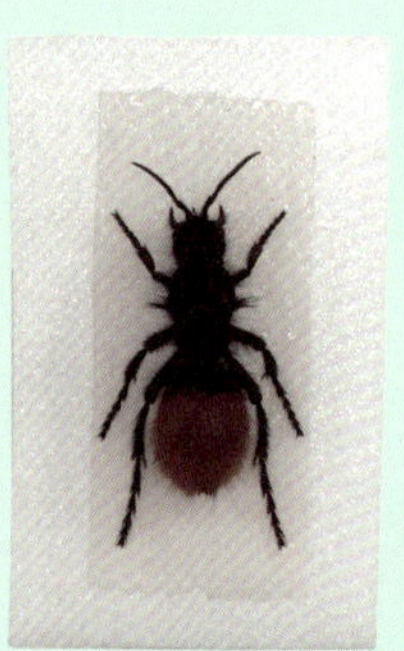
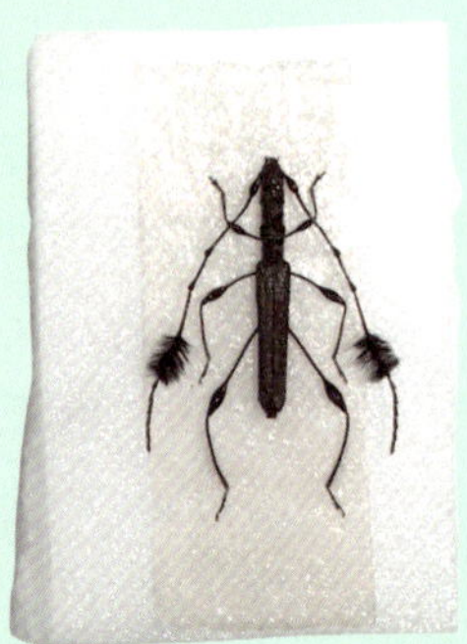

⑥マウント
Mount

展足針を外し、ラベルをつけて標本箱に収めることをマウントといいます。

The process of removing the pins, attaching labels, and placing the specimen in a specimen box is known as mounting.

針刺し標本
Pinned Specimens

最も一般的な標本手法。甲虫の場合右側後胸部に垂直に針を貫通させる。真ん中へ針を刺さないのは、針穴で失われた情報をもう片側で補完できるため。針を持てば裏側まで繊細な標本を安全に観察することができる。

This is the most common preservation method. For Coleoptera, a pin is vertically inserted through the right side of the metathorax. Inserting the pin on one side rather than the centre allows the preservation of information from the opposite side. Holding the pin enables safe observation of the delicate specimen, even on its underside.

よくない貼り付け方
Undesirable Mounting Techniques

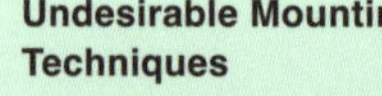

四角台紙・三角台紙双方のメリットを潰してしまう様な貼り付け方は避けたいところだ。また、プラスチック製の台紙を使用する人もいるが、以前真夏の車内に放置されたことでプラスチック台紙が熱変形を起こし壊滅的な被害を受けた標本を目にしたことがあり、あまりに悲惨な状況であったため個人的におすすめはできない。

It is best to avoid mounting methods that negate the advantages of both rectangular and triangular mounting boards. Additionally, plastic mounting boards are sometimes used, but if stored in unventilated rooms during hot summer weather, they may warp due to heat.

台紙貼り標本
Mounting on a Mounting Board

虫に針を刺さず、標本を接着した台紙へと針を刺す手法。主に小型の昆虫で使用する。台紙への標本の貼り付けは木工用ボンドか膠を使用する。

This method involves affixing the specimen to a mounting board and inserting a pin into the board rather than the insect itself. It is primarily used for smaller insects.

細貼り
Narrow Mounting

細長い四角台紙に虫を貼り付ける手法。針を刺すと変色しやすい虫や触角が極めて長い虫などに例外的に使用。

This technique uses a narrow rectangular mounting board to attach the specimen. It is used selectively for insects that easily discolour upon pinning or those with extremely long antennae.

三角台紙
Triangular Mounting Board

三角形or台形の台紙の先端に虫を貼り付け、虫に針を刺さずに裏側まで観察することができる手法。裏側に同定形質をもつ小型の虫はこの手法でマウントすることが望ましい。重い大型の昆虫には適さないほか、接着面が小さいため剥落しやすく、固定しにくいことから輸送による事故も起きやすいこと、ややスペースを大きく使うことがデメリット。

This method uses a triangular or trapezoidal mounting board, allowing the specimen to be affixed without inserting a pin directly into it, enabling observation of its underside. This technique is recommended for small insects that possess diagnostic characteristics on their undersides. However, it is unsuitable for heavy, large insects and has disadvantages such as a small adhesive area, making it prone to detachment and difficult to secure. Transport accidents are more likely, and it tends to occupy more space.

四角台紙
Rectangular Mounting Board

保存性が最も高い手法。虫体に触れず固定が可能なため破損しにくく輸送に強い。そのため、台紙内に虫が収まっていることが重要である。裏側を観察するには台紙から剥離させなくてはならないことがデメリット。

This is the most durable method, as the insect can be fixed without direct contact, making it resistant to damage and well-suited for transport. However, it is essential that the insect fits entirely within the board's area. A drawback is that to observe the specimen's underside, it must be removed from the board.

横貼り Side Mounting

四角台紙の裏側にポリフォームを貼り付け、そこに針を刺すことで横から観察できるようにしたもの。ツノゼミなど、立体的で横からのシルエットが特徴的な昆虫に向いている。

A piece of polyfoam is attached to the back of a rectangular mounting board, allowing a pin to be inserted so the specimen can be viewed from the side. This method is ideal for insects like Membracidae, which are three-dimensional and have characteristic silhouettes from the side.

ダブルピン Double Pin

現在は生産が行われていない特殊な針。メリットが三角台紙と概ね同じでありながら生産コストが高いため淘汰されてしまったと思われる。裏側が観察できるほか、撮影において威力を発揮する。

A special pin that is no longer manufactured. Although it offers similar advantages to the triangular mounting board, high production costs likely led to its discontinuation. This pin enables observation of the specimen's ventral side and is valued in photography.

修理

Repair

外れてしまったパーツは接着剤を用いて修理を行います。特に購入した標本は元から修理が行われており、軟化により接着が外れてしまうことも珍しくありません。適切な接着剤を用いて修理することが大切です。保存修復の観点から、剥離や除去が可能な木工用ボンド、膠、瞬間接着剤等を使用します。
シリコン系接着剤、強力なエポキシ樹脂やUVレジンなどは硬化後の剥離・除去が困難であるため使用は避けたほうが無難です。

For reattaching detached parts, adhesives are used. Especially with purchased specimens, it is not uncommon for parts already repaired to come off during softening. It is important to use appropriate adhesives for repairs, considering conservation and restoration. Wood glue, animal glue, and cyanoacrylate are suitable, as they allow for detachment or removal. Silicone adhesives, strong epoxy resins, and UV resins should generally be avoided due to the difficulty of removal once hardened.

熱整形 Heat shaping

アルコールランプ等で熱したピンセットを癖のついた跗節や触角などに押し当てることで修正を行う技法。テネラル(羽化後間もない体が固まりきっていない個体)等に有効。熱しすぎると焦げ跡がついたり、修復不能な破損を招くため注意が必要。
以前これを利用しクワガタムシの大顎などを変形させ、未記載種のように装って販売した標本商が存在した。

A technique involving pressing heated tweezers onto bent tarsi or antennae can be used for reshaping. This method is effective for teneral individuals whose bodies are not fully hardened shortly after emergence. However, overheating can cause scorch marks or irreparable damage, so caution is necessary. There have been cases where dealers used this method to reshape Lucanidae mandibles, presenting them as undescribed species for sale.

ラベル

Labels

標本には必ず採集場所が記されたラベルをつけるようにします。私はこの標本とラベルが針によってまとめられた一体感がとても好きです。種名は研究が進み変更されることがあるため、データラベルとは分けて作成することが一般的です。ラベルの紙は必ずケント紙等の中性紙を使用し、顔料インクで印刷します。油性ボールペンや光沢紙でラベルを作成すると経年で文字が消えたり滲んで解読不能になってしまうことがあるので注意が必要です。

Specimens must always be affixed with a label indicating the collection location. Given that species names may change as research progresses, it is common practice to create data labels separately. Labels must be made from acid-free paper. Care should be taken when creating labels with permanent markers or glossy paper, as the text can fade or bleed over time, rendering it unreadable.

MALAYSIA: N. Borneo,
Sabah, Near Keningau,
V/ 2006
Local Collector

西表島,上原,テドウ山,
JAPAN: Okinawa-ken,**[Iriomote-jima]**,
Yaeyama-gun, Taketomi-cho, Uehara,
Mt. Tedo-san, [H=126m],
24°22'57.72"N123°49'01.50"E
14/ X/ 2023, Keiki FUKUI leg.

データラベル
Data Label

基本的に採集場所・採集年・採集者名で構成される。GPS データや標高、採集方など、より詳しい情報を記す場合もある。飼育品の場合は採集年ではなく羽化日（em.=Emergence/emerged の略）を記す。（例：14 X 2023 em.）

Basically, it consists of the collection location, collection year, and collector's name. In some cases, more detailed information such as GPS data, altitude, and collection method may be included.
In the case of captive specimens, the emergence date (em.=Emergence/emerged) is listed instead of the collection year. (Example: 14 X 2023 em.)

Kazuho KOBAYASHI det. 2022

同定ラベル
Determination Label

種名のラベルはデータラベルとは分けて作製する。同定には間違いが起きたり、将来的に分類が変わる可能性があるためである。基本的に種名・同定者名・同定年で構成される。

Species name labels are created separately from data labels, because there is a possibility that identification errors may occur or that classification may change in the future. Basically, they consist of the species name, the name of the person who identified it, and the year of identification.

保存について

Preservation

photo: Hibiki Hoshito

標本は湿度や紫外線に弱いため、日本では気密性が高い“ドイツ型標本箱”に入れて保管します。UV カット仕様のものだとより安心です。
桐箱やボール紙製の標本箱も市販されていますが、カビや虫害の起きやすい日本では一時保管にとどめておくのが良いでしょう。一時保管には密閉できるタッパー容器も適しています。
日本で生産されている“ドイツ型標本箱”と、“ドイツ製の標本箱”は実は別物です。日本製のドイツ型標本箱は、ドイツで伝統的に使用されている標本箱の縦横の比率とサイズを模倣して作られています。あえて「ドイツ型（Duits）」と呼ぶのは、例えばフランスなどではドイツとは対照的に、より横に幅広い標本箱が普及しているためです。フランスでは現在でもドイツで一般的に使用される標本箱とは縦横の比率が異なる長方形の標本箱が一般的に使用されています。

Specimens are vulnerable to humidity and ultraviolet light, therefore, in Japan, they are stored in highly sealed "Deutsch boxes." It is preferable to use boxes with UV-filtering films on the glass surface of the lid for added protection.
While wooden, paper, or plastic specimen boxes are also commercially available, in Japan, where damage from mould and Dermestidae is more likely, their use should be limited to temporary storage. Plastic containers that can be sealed are also suitable for temporary storage.
It is important to note that the "Deutsch boxes" produced in Japan are different from those made in Germany. Japanese-made Deutsch boxes mimic the aspect ratio and size of traditional specimen boxes used in Germany. The term "Deutsch" is intentionally used, as in countries like France, the prevalent specimen boxes are broader and differ significantly in their proportions compared to the German ones. Currently, rectangular specimen boxes with different aspect ratios from those commonly used in Germany are prevalent in France.

防虫剤

Insect Repellents

標本箱には防虫剤を入れて定期的に交換し、カビや虫害を予防します。防虫剤に殺虫作用があるわけではないので、侵入を許さないことが何よりも大切です。

Insect repellents should be placed in specimen boxes and replaced regularly to prevent mould and insect damage. It is crucial to ensure that pests do not gain entry, as the repellents do not possess insecticidal properties.

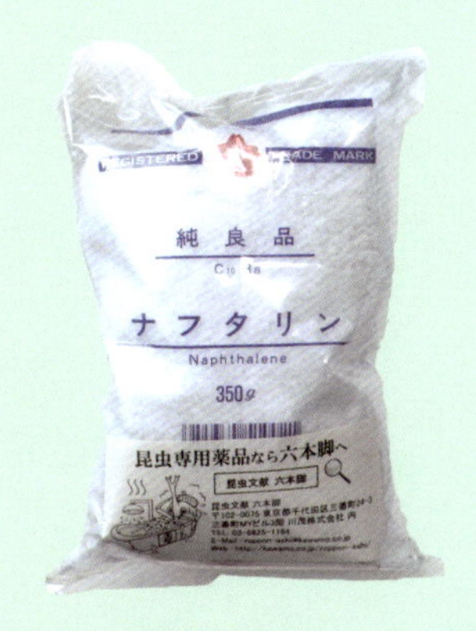

ナフタリン
Naphthalene

強力だが入手や扱いがやや難しい。大量に使用すると標本の表面に結晶が付着する場合がある。パラジクロルベンゼンとの併用は禁忌。

This is a powerful repellent, although it can be somewhat challenging to obtain and handle. When used in large quantities, it may leave crystalline deposits on the surface of specimens. It is contraindicated to use in conjunction with paradichlorobenzene.

パラジクロルベンゼン
Paradichlorobenzene

衣類用として入手が容易で安価。効果も強いが、油脂を溶かす性質があるため、多く使用すると脂を誘発させてしまう傾向がある。タッパーなども溶かして変形させてしまうため注意が必要。

This repellent is easily obtainable and inexpensive, often used for clothing. While it is effective, it has a tendency to dissolve fats when used excessively, which can exacerbate problems. Additionally, it may dissolve and deform plastic containers, so caution is warranted. There are also health concerns regarding its effects on humans.

ピレスロイド系防虫剤
Pyrethroid Insecticides

他の防虫剤との併用が可能だが、標本箱内では効き目がわかりにくい。また、抵抗力を持つ個体の発生が知られている。

These can be used alongside other insect repellents; however, their efficacy may be difficult to assess within the specimen box. There have been reports of the emergence of resistant populations.

クレオソート
Creosote

小瓶に脱脂綿などを詰め、染み込ませたものを固定して使用する。人体への有害性が高いため現在使用されることは少ない。

This is used by saturating cotton wool in a small vial and securing it in place. Due to its high toxicity to humans, its use is now rare.

カンファーオイル
Camphor Oil

楠の天然成分を抽出したもの。標本箱の木枠に染み込ませて使用する。

Extracted from the natural components of the camphor tree, this oil is used by soaking the wooden frames of specimen boxes.

ワサビ系防虫剤
Wasabi-based Insect Repellents

ワサビの成分を樹脂に練り込んだ防虫剤。交換時期がわかりにくいデメリットがあるが、防カビ効果は非常に高い。

This repellent is made by incorporating wasabi components into a resin. While the timing for replacement may be difficult to determine, it is highly effective against mould.

展足の歴史

小林一秀

昆虫標本そのもの自体の歴史や昆虫学成立までの由緒は、残された資料を辿ることによってその源流に迫ることができる。しかしながら「展足」の歴史となると、その正確な起源を特定することは困難を極める。これは昆虫の脚を整えるという作業は、科学的研究における利便性を求めた結果、あるいは標本をより美しく見せるためのアート性を高めるためワンステップとして、研究者やコレクターが編み出していった、言わば自然に発生したカルチャーであることに所以する。そして、それらの編み出されたスキルは、特に詳細に文体で記録されるわけでもなく、口伝によって代々のエントモロジスト〈昆虫愛好家／昆虫研究者〉たちに受け継がれてきたという特性ゆえにある。

研究、および趣味でのコレクション蒐集における展足で、最も重視されるスキルのうちのひとつと言えるシンメトリーを追求する美学は、古来より人間が感覚的に備えているものであるといってよい。石器時代の装飾品にもみられるほどに古くから存在している概念であることから、ただでさえ美しい昆虫の標本をシンメトリーを意識したポージングにすることでより美しく仕立てあげようとする試みは、エントモロジスト達によって築かれたカルチャーというより、むしろ人々の潜在意識中に存在するセンスである。したがって、昆虫標本の整形文化がいつごろから始まったのか、という歴史をたどることは不可能に近いであろう。

しかしながら、展足〈標本を正しく処理してポーズを整える〉という概念の重要性、またはその必要性が広く認識されるようになったのがいつごろの時代であるか、という点についてはいくつかの文献を紐解くことによってその発端をある程度たどることはできる。

昆虫学、ひいては近代動物分類学確立初期といって良い18世紀以前より、昆虫単体を被写体とし、その形態を写生したいわゆる「博物画」が残されている。昆虫単体の細密なイラストに関しては、古くはアルブレヒト・デューラー（Albrecht Dürer）のミヤマクワガタ（1505）などに代表される有名な例をはじめ（図1）、身近な自然産物を象徴するモチーフとして中世から多くの「美術的」なイラストが残されてきた。18世紀中期ごろまでに見られる昆虫をモチーフとしたイラストの多くは、昆虫の生前を彷彿とさせるような、躍動感のある生き生きとした姿を描いた作品がほとんどである。現代の学術的用途で使用される、昆虫を背面から俯瞰し、観察における利便性の観点から左右対称に脚を広げて描いたいわゆる「全体図（habitus image）」が増え始めたのは、18世紀後期頃からである。啓蒙思想の流行により近代的な博物学が成立していった18世紀には、昆虫の細部を科学的かつ、これまでよりもさらに合理的で写実的に描いた作品が数多く残された。例えば本書で取り扱ってきたような肉眼で観察可能な程度の大きさの甲虫に関しては、トーマス・ムフェット(Thomas Mouffet)、ジェイムズ・ペティヴァー(James Petiver)、カール・グスタフ・ヤブロンスキー(Carl Gustav Jablonsky)ら歴史に名を残す著名なナチュラリストたちが上梓した図鑑の図版に、数多く鮮烈なイラストと共に登場する。これらの作品以外にも、当時描かれた昆虫の細密画の例を挙げたら枚挙に暇がない。しかし、これらのいずれもそのほとんどが生前時の躍動的なポージングで描かれていることに対して、昆虫分類学への知見が深まってきた18世紀後半になると、フランス人昆虫学者であるギヨーム・アントワーヌ・オリヴィエ(Guillaume-Antoine Olivier)による大著「昆虫学、昆虫の自然史（Entomologie, ou histoire naturelle des Insectes)」(1789)に見られるように、基本的に昆虫標本を全て背面図で統一して紹介した文献が登場する（図2）。これは分類学が進むにつれてその形態的特徴の観察が重視され、従来の躍動感あるアート的要素を多分に含んだイラストより、全体を俯瞰することができ、脚や触角を含めた細部の観察が可能であるイラストの必要性が生じたことに由来するからであろう。また、それに伴い、基本的にシンメトリーの構造をもつ昆虫において、それを真上から俯瞰した際に、シンメトリーで描く必要性および芸術的価値観が生まれはじめた。

展足と一口にいえど、学術における形態観察の利便性を追求した左右対称な展足以外にも、昆虫の生前の姿を再現した「立体展足」という手法も存在する。この展足方法は前述のシンメトリーを追求したものとは大きく異なり、人間が持つ潜在的な左右対称を追求するという美意識から乖離し、昆虫が本来もつ行動学的・解剖学的なアプローチを取り入れたものである。この手法はある意味で左右対称展足よりも、より原始的な方法といえるかもしれない。歴史を紐解いてみれば、標本蒐集の文化は昆虫学あるいは分類学が成立する18世紀よりも古く、さらに数世紀も昔にさかのぼった時分に始まっている。当時のヨーロッパでは大航海時代を経て、海外の珍奇な品々がもたされ、それらの所持を社会的身分を反映させたステータスとして誇示する「脅威の部屋（Wunderkammerあるいはは Cabinet of curiosities)」とよばれる蒐集文化が上流階級の間で流行していた。海外で得られた珍妙不可思議な異形の動物たちの死骸は、上流階級の人々の知的好奇心を大きく揺さぶるのに充分たる魅力的な蒐集対象であった。そしてこれらの「死骸」を適切に処理し、インテリアとして客人に見せびらかすために、自室（キャビネット）にずらりと並べて展示するといった文化が、いわゆる「剥製（標本)」を作成する文化を生んだ。

たとえばオランダ人のフレデリック・レウシュ（Frederik Ruysch）は、17世紀後半に活躍した巨大なコレクション〈驚異の部屋〉の持ち主として有名な医師である。彼の蒐集の主な対象は人体をはじめとする医学的な標本であったが、昆虫も含めた動植物など多岐にわたった。レウシュは驚くべきことに、それらの標本をモチーフとした、インテリアとしての芸術標本作品を数多く作製した記録が残っている。彼の作製した作品を写生したいくつかのイラストの中には、昆虫をそのモチーフとして使用した形跡も見られる（Ruysch, 1710)。これは今でいう「立体展足」をおこなって作品を作り上げられたことは明確で、それらは左右対称を意識したものではなく、生前の動的なポージングを意識して作製されたものであった。昆虫の形態的特徴を紹介するために描かれた博物画は古くから数多く残されているものの、「標本・剥製」として死体を処理したことが明確にみてとれる作品は意外に少ない。レウシュのそれは「立体展足」が本人の意思でおこなわれたことを示す、最も古い例の一つであろう。

「左右対称展足」、それから「立体展足」、これらはいずれにおいても基本的な標本の作製手順という点に関していえばそれほど大きな違いはない。平面的、もしくは立体的に仕上げるために必要なスキルが異なるという点は除いて、死体の処理から作成完了までの流れにそこまで大きな手順の相違は見られないというのは、ここまで本書で紹介してきた通りである。それではその具体的な手法はいつごろから確立したのであろうか。

イギリスにおいて近代昆虫学が目覚ましいペースで発展を遂げた19世紀前半になると、昆虫学の父と称されるウィリアム・カービー（William Kirby）と、同じくイギリス人昆虫学者ウィリアム・スペンス（William Spence）による大著「昆虫学入門 Introduction to Entomology（1815-1826、全4巻)」が登場した。これはイギリスで最も古い昆虫学における分類学、生態学、解剖学、等の学術基礎をはじめ、採集方法や標本作成の方法についてまでも詳細が明記された、総合的な百科事典であり指南書である。この第4巻（1826年）には現在とほぼ変わらない基礎的な標本作成にまつわる手引きが記されており、本書でも紹介されているいくつかの基礎的な作成手法が既にこの頃にはある程度確立されていたことがわかる。

たとえば、展足作業にあたり軟化は不可欠であるが、その手法については蒸気を使用した軟化方法と、水分を直接体内に流し込む方法の2種類を解説している。前者は本書で紹介した湿潤軟化と同様の原理で、容器の中に砂と水を入れ、その上に昆虫を置いて蓋をし、2〜3日放置すると蒸発して発生した蒸気によって標本が軟化されるというものである。そして、後者においてはお湯の中に直接昆虫を投入し、水分を筋肉組織に吸い込ませることによって軟化を促進するという方法だ。いずれも今日まで使用されている方法であり、極めて原始的な手法とはいえ200年前には既に方法が確立されていたという点が興味深い。

さらに、昆虫のパーツを整形する理由として、形態観察における利便性を説きながらも「all of which tends much to the beauty of your specimens」（これらの手法はあなたの標本をより美しくさせる）と、展足によってその昆虫の美術的魅力を引き出すことが可能である点について記されている。この概念が文面として堂々と登場したという点において、この本を以って展足の源流と言ってよいかもしれない。

同じく19世紀前半の1839年、鱗翅目（蝶や蛾に代表される昆虫グループ）の研究者であり、王立昆虫学会の創立メンバーでもあったアビル・イングペン（Abel Ingpen）は「英国および外国産の昆虫における採集、飼育、保存方法の指南書（Instructions for collecting, rearing, and preserving British & foreign insects : also for collecting and preserving crustacea and shells）」（1839年改訂版）」において、昆虫標本の作成方法について鱗翅、甲虫ともにその手法と作成に用いる道具の詳細な紹介をカラーイラストを用いておこなった（図3・4）。この指南書では前述のカービー＆スペンス（1826）同様に、現代とほぼ変わらない標本作成に必要な道具やその手法が、当時既に完成しつつあることをうかがわせる。図4中には作成に必要な道具が記されているが、各種ピンセットや観察台など、現在のエントモロジストが使用を省略しがちな専門性の高い道具まで取り揃えられている。逆にこの中で現在と異なる点として挙げるとすれば、図4中・10番に描かれている針が道路のクランク状に曲がった標本針だ。これは図3中に見られるように符節を展足板から浮かずに押し付けるために使用するものとして当時は普及していたようだが、現在は使用されていない。この針の他に、三角形に切った厚紙（今で言ういわゆる三角台紙）を符節や触角の整形のために使用するという説明が見られる点など、これらから当時すでに「展足をする」という概念が明確に存在したことが見てとれる（図3）。また、これは前述のカービー＆スペンス（1826）にも既に同様の記述が見られるが、微小な昆虫を針刺し標本にすると破損の恐れがあることを考慮し、台紙張りにする文化もこのころには確立していた。

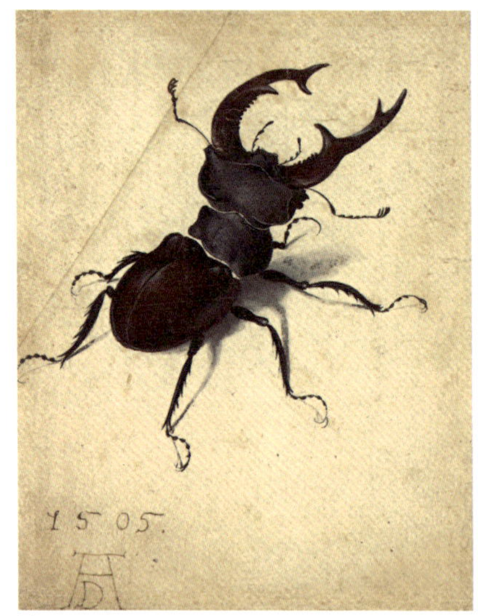

図1

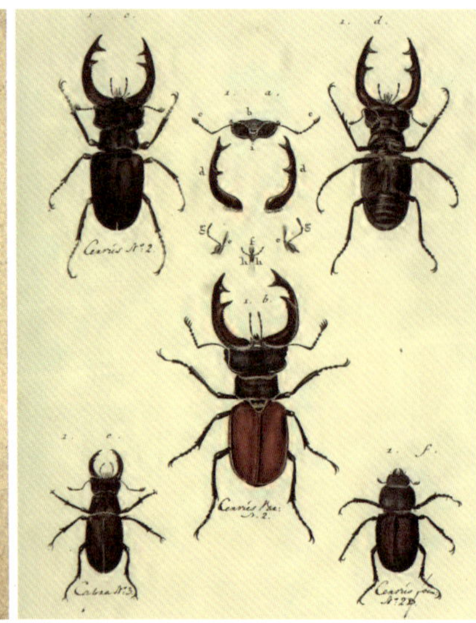

図2

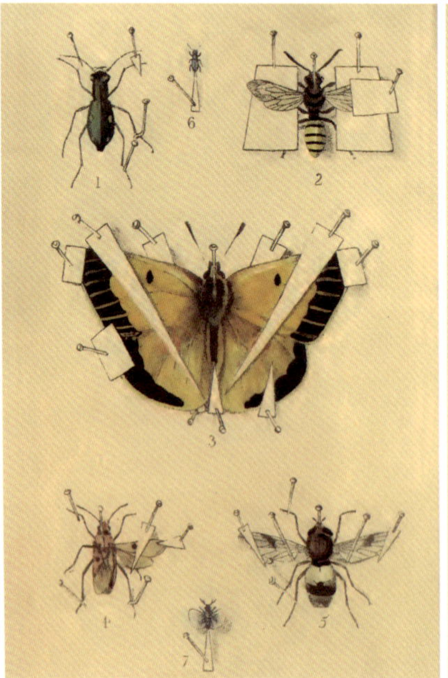

図3

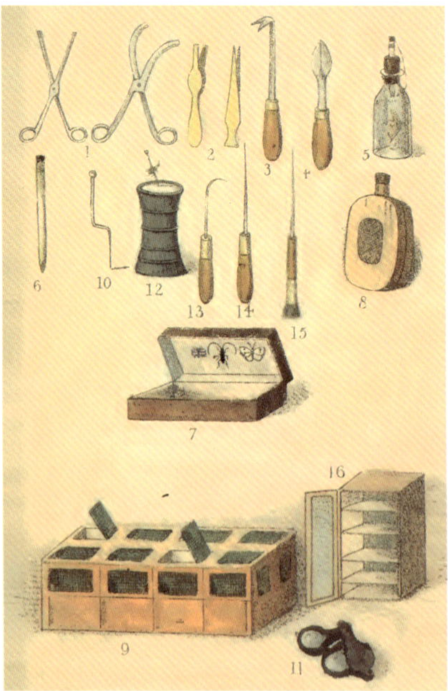

図4

図1 Albrecht Dürer, 1505. *Stag Beetle*. J. Paul Getty Museum.　図2 Guillaume-Antoine Olivier, 1789. *Entomologie, ou, Histoire naturelle des insects*. No. 1, Pl. 1.　図3 Abel Ingpen, 1839. *Instructions for collecting, rearing, and preserving British & foreign insects*. Pl 1.　図4 Abel Ingpen, 1839. *Instructions for collecting, rearing, and preserving British & foreign insects*. Pl. 3.　図5 Abel Ingpen, 1839. *Instructions for collecting, rearing, and preserving British & foreign insects*. Pl. 3.

References

Dürer, A., 1505. *Stag Beetle*. [illustration] (Watercolour on paper, 14.2 x 11.5 cm, J. Paul Getty Museum, Los Angeles, US).

Ingpen, A., 1839. *Instructions for collecting, rearing, and preserving British & foreign insects : also for collecting and preserving crustacea and shells, Second edition, with considerable corrections and additions*. London: William Smith. Pl. 1, 3.

Kirby, W., & Spence, W., 1822. *An introduction to entomology, or, Elements of the natural history of insects :with plates*, vol. 4. London: Longman, Hurst, Rees, Orme, and Brown. Pp. 536-536.

Oliver, G. A., 1789. *Entomologie, ou, Histoire naturelle des insectes : avec leurs caractères génériques et spécifiques, leur description, leur synonymie, et leurenluminée*, vol. 7. Paris: Imp. Baudouin. No. 1, Pl. 1.

Ruysch, F., 1710. *Frederici Ruischii anatom. & botanicesprofessoris ... Thesaurus animalium primus : cum figurisaeneis = Het eerste cabinet der dieren*. Amsterdam: Apud Joannem Wolters. Tab. 1

大英自然史博物館の甲虫コレクション

小林一秀

大英自然史博物館（ロンドン自然史博物館）は、ロンドンのサウスケンジントンに位置する世界で最も巨大かつ有名な自然史博物館のひとつである。その由緒は260年以上前の18世紀中期まで遡ることができ、量・質ともにこれまで数々の歴史上の著名な生物学者らも足を運んだほどの自他共に認める世界最高峰のコレクションを所蔵する。こと昆虫類の標本の充実度も群を抜いており、特に本書で取り扱った甲虫類の乾燥標本だけでも1,000万点以上、さらに毎年国内外からの新たな標本の寄贈や買取によってその規模は今なお拡大を続けている。

タイプ標本（分類学研究において使用される重要な標本）や、チャールズ・ダーウィン（Charles Robert Derwin）など歴史に名を刻んだナチュラリストたちの遺した貴重な標本は、主に荘厳な佇まいのカテドラル建築内の巨大なコレクションルームと呼ばれるバックヤードに眠っている。体育館を複数個並べて連結させたような巨大空間に、標本箱を収納するためのキャビネットが樹木のように整然と整列している。あまりの標本数の多さに、常駐のコレクションマネージメントを担当している学芸員でさえ、目当ての標本がある際は逐一博物館のデータベースでその所在地を確認しなければ標本のありかを特定できないほどのスケールだ。世界で最も広く使用されている言語「英語」を武器に、世界中の貴重な標本が連日次々に寄贈されている。特にイギリスは数々の国の旧宗主国であったことから、その結びつきの強い地域に産する昆虫の標本コレクションに関しては目を見張るものがある。

これだけの標本全てを博物館に勤務する学芸員だけで作成、ラベル付け、分類、整理することは実質的に不可能である。世界で最も複雑かつ多様性に満ちた生物である昆虫種の解明には国境を越えて世界中のスペシャリスト達による分担協力作業が不可欠であり、大英自然史博物館においても国内外から積極的に「ビジター」と呼ばれる外部研究者による収蔵品調査を迎え入れている。コレクションルームにはいたるところに机と椅子、それから実体顕微鏡が設置されており、こうした外部から訪れた研究者が気軽に調査に集中できるよう、万全の設備が整えられている。こうした博物館を訪れる研究者の手によって、膨大なコレクションは日々少しずつ整理されてゆき、標本資料としての役割を果たすことができるよう登録されていく。研究者は自身の持つ知識を活用してコレクションを研究、整理、メンテナンスをして博物館のために貢献する。また、博物館は研究者に貴重な資料を提供して科学の発展に尽くすという、まさに互いにとってメリットとなる相互関係が働いている。
このように多くの人々の協力により、今日までこれだけの巨大なコレクションを維持、管理することに成功しているのだ。

また、大英自然史博物館が世界最高峰の博物館として不動の地位を築いている要因の一つとして、現在のコレクション管理者のうちのトップであるマックス・バークレー（Max Barclay）主任学芸員や、松本圭太学芸員をはじめとした、学芸員らの専門知識はもちろん、学術研究以外での昆虫の普及活動への積極的な協力の姿勢などが間違いなくあげられるであろう。本書では大英自然史博物館のバックヤードに保管されている無数の標本の中から、特に汚れや状態の悪い標本を選び出し、正式な貸出許可を得てそれらを日本に持ち出し、展足を施した。一般的に「博物館所蔵の標本」と聞くと、はたしてどのようなイメージを抱くであろうか。博物館、それも世界最高峰の自然史博物館が所蔵する標本と聞くと、それらはいかにも完璧に整理され、非常に美しい状態で保管されている…もちろん展足も綺麗に施され、いつでも利用できるよう最高の管理がされている、そんな想像をする愛好家も多いのではないだろうか。しかしながらその実態とは、たしかにラベルの作成や種の同定作業などにおいては、博物館を支える多くの研究者によって完了されているものの、展足作業にいたっては所蔵標本のほとんどがおこなわれていない状態であると言っても過言ではない。おそらくクリーニングまで含めた、いわゆる「綺麗に」展足がしっかりとされている標本は、数千万点のコレクションのうちわずか1％にも満たないのではないだろうか。「世界最高峰の博物館が所

有する標本のほどんどが展足がされていない状態」と聞くと、実際に博物館のコレクションを目の当たりにしたことがない人間にとってはややセンセーショナルな事実と捉えられるかもしれない。しかし博物館の標本保管、所蔵における本懐は「研究のため」であり、標本に美術的価値を与えることではない。端的に言えば、仮に何かの理由で展足作業の必要性が生じた場合は、その都度必要とする人間が展足すればよいだけの話である。また、逆説的に、展足が行われていないのは、絶えず拡張を続ける標本コレクションの全てを展足するための膨大な時間、および学芸員の確保など、現実的に不可能であるからだとも言える。しかしながら展足は研究の際により容易に昆虫の形態的特徴を可視化するため、または展示においてその潜在的な美しさを引き出すため、極めて有効な作業であることには間違いない。その啓蒙普及活動的役割を担う本書の作成を、バークレー主任学芸員をはじめとする博物館の学芸員らは快く受け入れ、サポートしてくれた。この研究の面ではもちろんのこと、昆虫というカルチャーの普及という全ナチュラリストに課せられた命題ともいえる活動に積極的に関与するフレキシビリティこそ、大英自然史博物館の強みであり、最強の博物館たる所以であろう。

大英自然史博物館の輝かしい歴史は今日始まったものではない。大英自然史博物館は元々はロンドンのブルームズベリーの地に位置していた大英博物館（The British Museum）がその前身にあたる。1759年に一般公開が始まった大英博物館は、開館にあたりイギリスの医師・博物学者ハンス・スローン（Hans Sloane）が生前に収集した膨大なコレクションを基盤にスタートした。驚異の部屋（Cabinet of curiosities）と呼ばれた当時の蒐集ブームの最中においてスローンは、親友であり昆虫類の標本の蒐集に並々ならぬ情熱を燃やした同じくイギリス出身のジェイムズ・ペティヴァー（James Petiver）から、彼の死後に£4,000を支払い巨大な昆虫標本コレクションを相続した。こうして飛行機どころか、海外へ出向くのにも命がけの航海が必要であった時代にもかかわらず、スローンが遺した5000点以上の標本をもとに大英博物館はその後も徐々にコレクションを強化していった。19世紀に至るとイギリスはビクトリア朝の時代を迎え、これはまさに歴史上イギリスの黄金期ともいえる時代であった。文化的、そして科学的にも他の追随を許さない確固たる世界最強の強豪国へと上り詰め、イギリス国内での産業活動の発展や、航海技術の目覚ましい発展により、これまで以上に海外の未開の地や植民地支配を得た地域から様々な昆虫をはじめとする珍奇な標本が入手できる時代へと突入した。

現在では最も基礎的な学問と言える生物分類学であるが、18世紀に分類学の父・カルロス・リナエウス（Carolus Linnaeus）が二名法を提唱して以降長年にわたって「学問の王道」と称され人気を博した。そんな中、当然博物館も新たなる種の蒐集に注力し、結果として様々な地域の昆虫を入手することに成功した。また、大英博物館のコレクション充実の要因の一部として、昆虫のベイツ型擬態で知られるヘンリー・ウォルター・ベイツ（Henry Walter Bates）と、自然選択説で知られるアルフレッド・ラッセル・ウォレス（Alfred Russel Wallece）の活躍もあった。彼らは共に良き友人関係にあり、採集した標本を大英博物館に買い取ってもらうという約束をもとに1848年に南米のアマゾンへ向かった。進化論で有名なかのチャールズ・ダーウィンも昆虫には並々ならぬ関心を寄せており、自身が採集した標本の一部も最終的には大英博物館に収蔵された。

しかしながら日々拡大を続ける標本コレクションに収蔵施設のキャパシティが追い付かず、当時自然史以外にも総合的なコレクションの収蔵の役割を担っていた大英博物館は、とうとう限界を迎えてしまう。標本は床に山積みにされ、収蔵施設そのものの老朽化なども目立つことから、ついに当時の自然史部門長を務めていたリチャード・オーウェン（Richard Owen）によって独立した自然史部門だけで構成される自然史博物館建築の必要性が提唱された。1881年には満を持して前述の巨大なカテドラル建築の荘厳たる博物館が開館し、今日までその姿をとどめている。

1. 大英自然史博物館を象徴する荘厳な佇まいのヒンツホール（Hintze Hall）。2. 博物館全景。巨大なロマネスク様式建築。3. バックヤードの通路に壁のように立ち並ぶ、標本箱を収蔵したキャビネット。4. 壁際の通路には外部からやってきた研究者が自由に使えるデスクが複数ある。5. バックヤードの拡張工事によって発見されたというヨーロッパミヤマクワガタのレリーフ。6. ニジダイコクコガネ（Phanaeus）の仲間を収めた標本箱。7. まるで鉱物のような圧倒的な存在感を放つ南米原産のランキフェルニジダイコクコガネ（Coprophanaeus lancifer）。8. 歴史的著名人に献名された甲虫種を集めた展示標本箱。

TENSOKU
-Insect preparation-

Author
KEIKI FUKUI
KAZUHO KOBAYASHI

Photographer
KOJI SAKAI

Translation
Carim Nahaboo (www.carimnahaboo.com)

Special Thanks
The Natural History Museum, London
Max Barclay (NHML)
Keita Matsumoto (NHML)
Michael Geiser (NHML)
丸山 宗利（九州大学総合研究博物館）
藤岡 昌介
飯島 和彦（有限会社むし社）
佐藤 仁
前田 健（株式会社インセクエスト）
川上 暁
丸山 圭太
鈴木 賢紀
法師人 響

STRAIGHT

Editor
Shogo Kawabata

Art director
Hideaki Komiyama

straight-collective.com

福井敬貴　KEIKI FUKUI

2017年多摩美術大学美術学部彫刻学科卒業。2019年同大学院修士課程彫刻専攻修了。主に昆虫標本を用いた作品や虫をモチーフとした立体作品を制作する他、標本の制作依頼や展示会・書籍等への協力、3DCGモデルの造形・原型監修など分野を横断した活動を行っている。共著書に「とんでもない甲虫(幻冬舎)」「世界で一番美しい甲虫図鑑(誠文堂新光社)」
X : @fukuinsect

小林一秀　KAZUHO KOBAYASHI

甲虫標本商花滝代表。2020年英国Nottingham Trent University, Animal Biology学科卒業。コガネムシ上科の新種の発表など、精力的に研究活動を行っている。専門学校で講師としても従事し、また昆虫に関する展示会やTV番組作成の協力等も行うなど、様々な活動に取り組んでいる。
https://www.cogazo.com

撮影
堺 浩二　KOJI SAKAI

Still Life Photographerとして自身のスタジオ「Relight」でブランドビジュアル制作を行う傍、"Photographic Sculptor"としてアート作品の制作も行う。有機物から人工物まで様々な被写体に向き合い、それぞれが秘める美しい瞬間を追い求め、「永遠の現象」として写真に封じ込める。
https://kojisakai.com

STRAIGHT

大航海時代後に世界中から珍品を集められた博物陳列室「好奇心の部屋(Cabinet of Curiosities)」を現代の感性で蘇らせることをテーマに、森羅万象への飽くなき好奇心を呼び覚ますヴィジュアルブックを作るパブリッシャー。
https://www.straight-collective.com

昆虫展足の世界

2024年12月10日　第1刷発行

著者	福井敬貴／小林一秀
発行者	川端正吾
発行所	STRAIGHT（shikumi有限会社内） 〒150-0002 東京都渋谷区渋谷1-3-18 C302 TEL 050-7114-0718
発売	株式会社メディアパル（共同出版者・流通責任者） 〒162-8710 東京都新宿区東五軒町6-24 TEL 03-5261-1171
印刷・製本	シナノ印刷株式会社

ISBN 978-4-8021-3498-9